NAVAL
AIR POWER

NAVAL AIR POWER

MICHAEL TAYLOR

HAMLYN

Half-title Tupolev Tu-22M Backfire
(courtesy of the Swedish Air Force).

Title spread A US super carrier prepares to launch
its Tomcats on an air defence sortie
(courtesy of Brian M. Service).

Published 1986 by
Hamlyn Publishing,
Bridge House, London Road,
Twickenham, Middlesex

ISBN 0 600 50115 9

Printed and bound by Graficromo s.a., Cordoba, Spain

Contents

Introduction

Because of the nature of aviation history, arbitrary decisions have to be made regarding the subjects to cover in a particular book and, often more difficult, the historical starting point. Do myths and fantasies have any part to play in the annals of history, beyond indicating that man considered the art of artificial flight many centuries before it became a practical reality?

At first a book about naval air power appears to have a clear starting point – one often chosen and in most respects entirely satisfactory. This is the work of the Frenchman, Clément Adler, and thereafter of the American, Eugene Ely, with his epic flights from and on to temporary platforms constructed on-board US Navy warships. True, these represent the origins of aircraft carrier aviation as we know it today, but does naval air power predate the late 19th century? The answer is that it does, as detailed in the first chapter of this book.

Naval Air Power is, however, a book mainly about modern trends, the operations and tactics of aircraft used in a maritime role and, of course, the equipment in terms of both aircraft carriers and the aircraft they operate. It is for this reason that only the first chapter is devoted to the historical aspect of naval air power.

This book attempts to answer why naval air power is such a dominant force within military aviation. It is not just that a modern aircraft carrier offers an operating nation a fully self-sustaining and mobile floating air base, from which it can apply political power by threat of action or can mount devastating air attacks while protected by on-board aircraft from surprise enemy air, surface and undersea attack. Equally, shore- and sea-based aircraft can often prove the most effective way of locating, tracking and dealing with the sub-surface and air threats which carry the awesome power of atomic weapons in strategic roles.

The author would like to thank all the individuals, companies and organizations from around the world whose assistance and generosity in supplying information, technical data and illustrations have contributed to this book. Special thanks go to the Grumman Aerospace Corporation, British Aerospace, Roy Braybrook, Steve Broadbent, Bob Hutchinson and Mike Spick.

◀
A fantastic aerial exploration vessel named Minerva, designed by Professor Robertson and engraved at the Ordnance Survey Office in Southampton in 1864. Armed with cannon, fully self sustaining, and with its on-board aircraft in the form of balloons and steerable parachutes, it was clearly conceived to provide naval air power.

▷
The Harrier V/STOL aircraft – modern exponent of naval air power.

Chapter 1
And so it begins...

More than a century before the French Montgolfier brothers conquered the air with hot-air balloons, at last enabling man to enjoy sustained flight, an attempt had been made to design and build a 'flying boat' that was both serious and scientific. Using von Guericke's newly invented air pump, Jesuit priest Francesco de Lana-Terzi reasoned that he could construct extremely thin copper globes from which he could withdraw the air by using a pump, causing them to be lighter than the displaced air in the same way as a boat is able to float in water. By attaching these globes to a boat hull, a flying machine of awesome capability would result. There was, however, one major flaw in his theory. If the globes were made sufficiently thin for lightness, atmospheric pressure simply crushed them as the pump went to work extracting the air. Clearly de Lana had met an insuperable problem and he had to give up his bid to construct a full-size craft.

Interesting though de Lana's craft was, has it any place in naval air power history? Having failed to fly, de Lana wrote in 1670 that 'God would not suffer such an invention', prophesying how a craft such as his might one day be manoeuvred over buildings and ships at sea to drop 'artificial Fire-works and Fire-balls'. Although he was accurate in his prognosis of how flying machines could be used, the earlier failure had not stopped him from attempting artificial flight and it did not stop others;

and, surprisingly, there is a spurious connection between his writings and what the author considers to be the true beginning of naval air power.

At various times from 1780, the British were repelled in their advances in India by an enemy armed with a devastating weapon, the iron-clad artillery rocket. The first occasion on which it was used was at Guntur and it was still affecting progress nine years later. Clearly the artillery rocket was a weapon of the future and Britain began her own experiments at the Royal Laboratory at Woolwich, under Colonel William Congreve. By 1805 sufficient progress had been made for tests to begin of his developed rocket. In one of the lesser known engagements of the Napoleonic War, on 8 October 1806, the Royal Navy sailed to Boulogne to attack the town and French Navy ships, the mighty British fleet including 24 ships armed with Congreve rockets. These rockets caused much damage and in 1807 the Royal Navy bombarded Copenhagen with a recorded 25,000 rockets, having a range of about 2 miles (3.25 km).

These attacks were just the beginning. The Royal Navy used similar rockets to mount offshore bombardments for decades afterwards, including during the famed bombardment of Algiers on 26 August 1816 and the attacks on Frederikshamn during the Baltic campaign on 21 July 1855. The early use of rockets from naval vessels was clearly the precursor of today's ship-launched missiles.

Preceding pages

Ely flies off USS Pennsylvania, *18 January 1911.*

▼

Bombardment of Algiers using rockets, 1816.

Enter Ader and Langley

During the last quarter of the 19th century it appeared that a powered aeroplane using steam engines would prove that sustained heavier-than-air flight was possible. Hop flights were then taking place in several countries,

Langley's full-size Aerodrome *before launch.*

although France was leading the way, as it was also in the development of the steerable airship (dirigible). On 9 October 1890 a Frenchman, Clément Ader, made the world's first 'hop' in a powered aeroplane from level ground, his steam-powered *Eole* monoplane looking like a huge mechanical bat with the pilot in a fabric-covered fuselage nacelle and the engine driving a primitive paddle-like tractor propeller. In February of the following year Ader was put under contract by the French Minister of War to build a two-seater for military purposes, requirements including the ability to carry a light bomb load. Although it was built, this machine was not successful and it crashed at Satorg on 14 October 1891. Official backing was withdrawn. Despite being unable to fulfil the contract, Ader's thoughts on possible applications for aircraft went far beyond the technology of the day. In *L'Aviation Militaire*, published in 1909, he gave an extremely accurate description of a modern-day aircraft carrier, detailing a flight-deck free from obstacles, a lift to a below-deck hangar for maintenance, and the vessel's speed equal to or surpassing that of a cruiser. Ader's concept is historically important, but at the time it was written the aeroplane itself had already become a reality, not least through the invention of the petrol internal combustion engine and the achievements of the Wright brothers. However, a US government-backed aeroplane had nearly flown before they took to the air.

In 1891 the American physicist, Samuel Pierpont Langley, began experiments in which he eventually catapulted large model aeroplanes from a houseboat on the Potomac River. In June 1901 Langley launched a quarter-scale unmanned model from the houseboat and recorded the first sustained level flight by a petrol-fuelled aeroplane. On 7 October 1903 the full-size Langley aircraft, like the models named *Aerodrome*, was ready for its first trial launch from the houseboat, piloted by the engine designer, Charles M. Manly. Catapulted, it caught the launcher and dropped into the Potomac. A second attempt on 8 December again resulted in the launcher being fouled and Manly took another soaking. On 17 December Orville Wright piloted his aeroplane on its first twelve-second flight, and support for Langley was subsequently withdrawn.

Second to the Army

Representatives of the US Navy were present during the 1908 US Army trials of a Wright biplane at Fort Myer, Virginia, and must have been impressed by what they saw, with the exception of the crash that killed an Army officer and badly injured Orville Wright. While trials were under way, Wilbur Wright had journeyed to Europe and gave flying demonstrations. These flights were sensational. While at Camp d'Auvours in France on 21 September 1908, he flew 41.3 miles (66.5 km), outclassing other aeroplanes in existence to such an extent that most of Europe's sport flyers were clamouring for the chance to purchase examples. In late August 1909 at Reims, the site of the first ever international aeroplane meeting, three Wright biplanes took part along with many others of the world's best aircraft (which, incidentally, were mainly

French). So impressive was the flying, that the US Naval Attaché to Paris recommended that the US Navy should convert a battleship and build other vessels to carry and launch aircraft, a suggestion that was dismissed by the Department of the Navy.

The *World* and the *Daily Mail*

The British newspaper, the *Daily Mail*, was foremost in aviation sponsorship at this time and it was in order to meet the challenge of one of its prizes that an Englishman, Hubert Latham, and a Frenchman, Louis Blériot, risked all in attempts to fly across the English Channel by aeroplane. Blériot managed this feat on 25 July 1909. In the USA, the Wright brothers' supremacy of the air was, by 1910, no longer total; the years the brothers had spent in promoting their aircraft and not flying had allowed others to catch up technically. The first American to fly in a powered aeroplane after the Wrights was Glenn Curtiss, whose *June Bug* had flown initially on 20 June 1908 and whose subsequent aircraft proved excellent. In May 1910 Curtiss won a substantial prize offered by the New York *World* newspaper, and he subsequently took the US Navy into the age of the aeroplane.

Elsewhere, on 19 January 1910 Lt Paul Beck had demonstrated that aeroplanes could be used as bombers, when he released sandbags over Los Angeles from an aircraft piloted by the sport flier Louis Paulhan. The

World, realizing the offensive possibilities of aeroplanes, decided to stage a dramatic demonstration. Having arranged the setting-out of markers on Lake Keuka in the shape of a battleship, Curtiss was requested to fly over and release lead pipe 'dummy bombs' in a simulated attack. This occurred on 30 June 1910, by which time Frenchman Henri Fabre had become the first man to fly an aeroplane from water (28 March 1910).

Enter Ely

A race to become the first to fly an aeroplane from a ship now began, which had the effect of forcing US Navy pace. The *World* newspaper and the commercial Hamburg-Amerika Steamship Line planned to upstage everyone by launching Canadian pilot J. A. D. McCurdy in an aeroplane from a liner, the purpose being to see whether such flights could improve existing mail services. Not to be beaten to this honour, the US Navy hurriedly decided to fix a wooden platform on the cruiser USS *Birmingham* and fly off the first aeroplane.

Having managed, by a stroke of luck, to gain the assistance of show pilot Eugene Ely for the experimental off-ship flight (Ely worked for Curtiss and thus had access to one of the best machines in the country), on 14 November 1910 Ely took off from the 83 foot (25 m) platform over the bow of the cruiser which had weighed anchor in Hampton Roads, Virginia. The plan had been to take off while the vessel steamed at 20 knots, but Ely had been impatient to begin. In the event his impatience nearly caused disaster, the Curtiss touching the water as Ely struggled to gain lift. Although the propeller had been

The US Navy's first aeroplane was the Curtiss A-1 Triad hydroaeroplane.

▲
S.38 ready for launch once HMS Hibernia is under way. Note that the platform is over the forward gun turret.

▲ ▶
Short Folder S.64, Royal Navy serial 82, moored at Spithead for the Naval Review of July 1914.

damaged, Ely landed safely at Willoughby Spit, 2½ miles (4 km) from the cruiser. But what of McCurdy? All had been readied for his take-off by 12 November (already after a postponement) but the aeroplane propeller was damaged and a further delay was thus incurred.

By now a Royal Navy officer had learned to fly on the Short S.26, Lt G. C. Colmore being awarded his pilot's certificate on 21 June 1910 and thus becoming the first naval officer in the world to qualify as a pilot. On 20 August 1910 a US Army pilot, Lt Jacob Earl Fickel, fired the first gun from an aeroplane (a rifle from a Curtiss) and on 7 January 1911 Lt Myron Sidney Crissy and Philip O. Parmalee crewed a Wright biplane from which the first live bombs were dropped, proving that aeroplanes could be armed successfully.

On 18 January 1911, Ely made the first landing on to a ship, at San Francisco Bay, the designated vessel (USS *Pennsylvania*) having been fitted with a 119 ft 4 in. (36 m) platform. After this feat the Captain, C. F. Pond, reputedly said: 'this is the most important landing of a bird since the dove flew back to the Ark'. How right he was. After lunching with Captain Pond, Ely took off from the armoured cruiser. A few days later, on 26 January 1911, Glenn Curtiss himself flew one of his 'hydroaeroplanes' to San Diego Harbor, where he performed the first water landing, taxi and take-off. On 1 July of the same year, the US Navy took into service a Curtiss A-1 hydroaeroplane, its first aeroplane.

By October 1911 the aeroplane was at war for the first time, Italian Air Flotilla aircraft based at Tripoli first undertaking reconnaissance flights over Turkish pos-

itions at Azizia and, from November, attacking with Cipelli grenades. More importantly to naval aviation, in December 1911 the Royal Navy's Lt Charles Rumney Samson flew secretly from a platform on HMS *Africa* in a Short biplane, while the battleship was anchored in Sheerness Harbour. Officially he performed the first British take-off from a ship on 10 January the following year, the Short S.38 again leaving the *Africa*. On 9 May 1912 Samson flew a Short from the forecastle platform on the battleship HMS *Hibernia* while the vessel steamed at over 10 knots, the first time an aeroplane had flown from a moving ship. This was undoubtedly the high point of the May 1912 Naval Review at Portland. In October that year Samson was appointed the commanding officer of the Naval Wing, Royal Flying Corps.

The first true aircraft carriers

Ships specially built or converted to carry aeroplanes date from 1912, when the French Navy converted the torpedo boat *Foudre* to carry seaplanes. However, half a century earlier, General McClellan's Army of the Potomac had commissioned a converted coal barge to carry and tow observation hydrogen balloons during the American Civil War. This vessel was used from November 1861 and helped the observation of Confederate forces.

In late 1912 the Royal Navy put into service a converted cruiser named HMS *Hermes* which was intended to carry two seaplanes, launching them from wheeled trollies. From 1913 onwards, it was assigned Short biplanes, known as Folders because of their revolutionary manually folding wings, although one had been completed without wing folding. This fixed-wing aircraft and *Hermes* took part in the Royal Navy's first-ever fleet manoeuvres involving aeroplanes in July 1913, and was joined by another from the spring of 1914. The refinement of folding-wing aircraft greatly enhanced the ability of ships to cope with the large dimensions of aeroplanes, and it was an important step in the development of the fully workable aircraft carrier in its present-day form.

Sopwith Pup

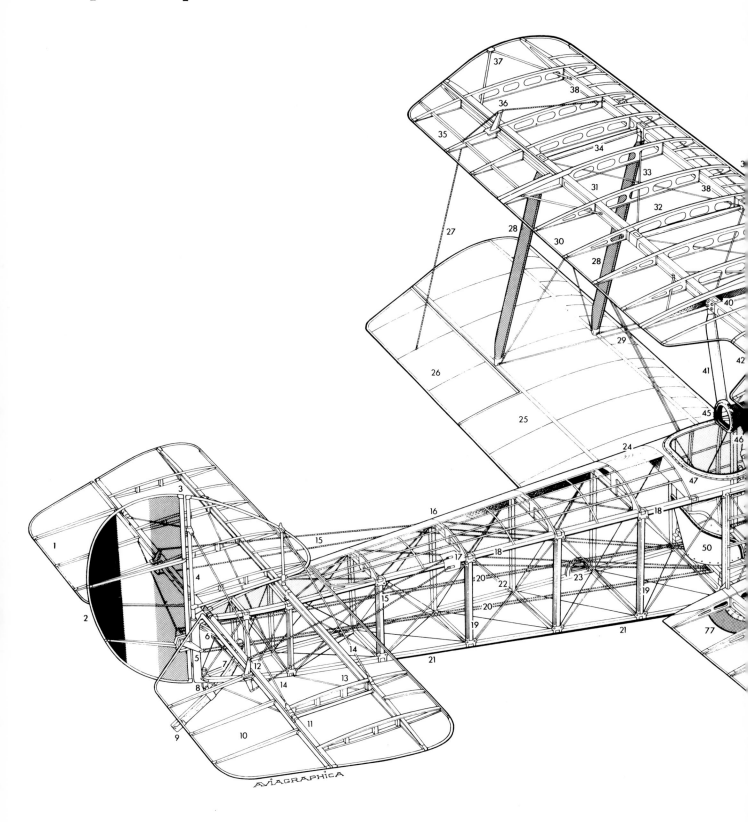

1 Fabric-covered port elevator
2 Fabric-covered rudder
3 Light tubular steel fin and rudder construction
4 Rudder post
5 Rudder operating crank
6 Tailskid elastic cord shock absorber
7 Wooden tailskid
8 Tailskid hinge mounting
9 Steel tailskid shoe
10 Starboard elevator
11 Elevator hinge bar
12 Elevator operating crank
13 Tailplane rib construction
14 Tailplane bracing wires
15 Elevator cables
16 Fabric-covered rear fuselage top decking
17 Elevator cable guide panel
18 Top longeron
19 Vertical spacers

20 Rudder cables
21 Bottom longeron
22 Fuselage cross bracing
23 Entry step, port
24 Plywood top decking
25 Port lower mainplane fabric
26 Port lower aileron
27 Aileron connection cable
28 Interplane struts
29 Diagonal wire bracing
30 Light steel tube trailing edge
31 Rear spar
32 Wing ribs
33 Diagonal bracing wires
34 Spar bracing strut
35 Port upper aileron
36 Aileron operating crank
37 Wing-tip diagonal bracing frame
38 Front spar
39 Leading-edge construction
40 Port upper wing spar joints
41 Wing centre-section struts
42 Trailing-edge cut-out
43 Fixed synchronized 0.303 in. (7.7 mm) Vickers machine gun
44 Gun synchronizing drive
45 Padded pilot's face guard
46 Gun cocking lever
47 Padded cockpit coaming
48 Instrument panel
49 Control column
50 Pilot's seat
51 Cartridge ejector chute

52 Rudder bar
53 Ammunition tank
54 Ammunition feed chute
55 Petrol tank
56 Engine bearer frame
57 Engine bulkhead
58 80 h.p. Le Rhône rotary engine
59 Aluminium engine cowling
60 Propeller hub
61 Two-bladed wooden propeller
62 Starboard upper wing ribs
63 Leading-edge stiffeners
64 Front spar
65 Spar bracing strut
66 Wing internal wire bracing
67 Wing-tip diagonal bracing frame
68 Aileron balance cable
69 Aileron operating crank
70 Starboard upper aileron
71 Interplane bracing wires
72 Light steel tube trailing edge
73 Engine cooling air duct
74 Fabric-covered fuselage framework
75 Footboards
76 Lower wing/fuselage attachment rib
77 Port mainwheel
78 Undercarriage vee strut
79 Axle beam
80 Half axle pivot fixing
81 Undercarriage bracing wires
82 Starboard mainwheel
83 Axle hub
84 Wing internal bracing wires
85 Interplane struts
86 Diagonal bracing wires
87 Spar bracing strut
88 Lower wing ribs
89 Aileron connecting cable
90 Starboard lower aileron
91 Aileron operating crank
92 Wing-tip diagonal bracing frame
93 Light steel tube wing-tip

◄ A drawing from Lowe's collection of a balloon ascending from the world's first aircraft carrier, the USS George Washington Parke Custis, to undertake a reconnaissance of the blockade near Budd's Ferry, November 1861.

▷ Because of their marginal performance while carrying a 14-in. Whitehead torpedo, most Short Type 184s and Improved 184s (as shown here) were usually armed with up to 520 lb (236 kg) of bombs suspended from an underfuselage rack.

Although the US Navy and the French Navy had gained important 'firsts' in the development of aircraft-carrying ships, it was Britain that led the field thereafter. At the instigation of Winston Churchill and Sqn Cdr A. Longmore, a Short Folder was modified to accept a standard 14-in. Navy torpedo. With such a weapon fitted, on 27 July 1914 Short test-pilot Gordon Bell made the first ever standard torpedo drop from an aeroplane, at Calshot, and the next day Longmore did the same.

Air power goes to war

At the outbreak of the First World War in 1914, the British government immediately sent most of the Army's aircraft to France, together with a squadron of Royal Naval Air Service seaplanes, and air defence of the British Isles was therefore left to the Royal Navy.

The immediate priority was to get more seaplanes to sea, and to this end the Admiralty requisitioned three cross-Channel steamers, the SS *Empress*, *Engadine* and *Riviera*, in order to convert them for aircraft operation. However, their endurance proved limited, and the old record-breaking Cunard liner, *Campania*, which had been sold for scrap and was lying derelict, was picked to accompany the battleships of the Grand Fleet. She can thus justifiably be credited with being the first fleet carrier.

The conversions took several months, but the RNAS squadron in France was soon involved in action; its pilots carried out valuable scouting and reconnaissance sorties, returning with information about German troop concentrations. Not content with this, the Admiralty organized a bombing raid on the Zeppelin airship sheds at Düsseldorf, on 22 September 1914, but the raid was a failure because of bad weather and the unreliability of the bombs used. However, the next raid on 8 October not only destroyed a shed but also the airship inside it.

Spurred on by this achievement, the Royal Navy set its sights much higher, and the result was the first use of naval air power as we understand it, with a carrier force being deployed offshore and sending its aircraft to attack a target on land.

In this attack, on Christmas Day 1914, the Royal Navy aimed across the North Sea at the German base of Cuxhaven. The three modified cross-Channel ships were used, each carrying three seaplanes, in conjunction with a force of light cruisers, destroyers and submarines, and this fleet stood off Heligoland while the seaplanes headed for Cuxhaven. Unfortunately, the airship sheds were not at Cuxhaven but to the south at Nordholz, and only one aircraft found them, its bombs falling into trees. The returning pilots alighted on the sea and were recovered by a submarine which had surfaced; the main fleet had withdrawn earlier because of the danger of attack from German ships so close to their home base.

Although the Cuxhaven raid had not been effective, the Admiralty had not been discouraged about the principle of using seaplane carriers and they went ahead with the conversion of a further three ships, the steamers *Ben-my-Chree*, *Manxman* and *Vindex*. At this time the new carrier *Ark Royal* joined the Fleet and was immediately sent to the Mediterranean to help in the Dardanelles operation.

The *Ark Royal* and her eight seaplanes were given the job of reconnaissance and spotting for the heavy guns of the accompanying battleships, a supporting role which was ascribed to aircraft carriers until as late as 1940 (only five years later the battleship had been rendered obsolete by the awesome power and flexibility of the aircraft carrier). The ungainly seaplanes flew to their absolute limit, and soon Wing Commander Samson (who commanded the early RNAS squadron in France) took over a squadron of lighter landplanes in the form of Nieuport Scouts. These achieved great nuisance value by attacking Turkish positions and disrupting road transport. Because the *Ark Royal* was rather slow she was vulnerable to the German U-Boat menace, so in mid-1915 she was replaced by *Ben-my-Chree*, which carried two Short seaplanes. These were each capable of carrying one 14 in. torpedo, and they were used in anger for the first time on 12 August 1915 in the Sea of Marmara. Flight Commander C. H. Edmonds managed to sink a 5000 ton Turkish supply vessel with a torpedo, but this 'kill' was hotly disputed by a British submarine, which claimed to have torpedoed the same vessel. However, there was no dispute over a steamer and a tug sunk by Edmonds and a colleague five days later. These ships were the first ever to be sunk by an air-launched torpedo, but the credit for the first ship to be sunk from the air goes to the Japanese, whose Farman seaplanes bombed a German mine-layer off China in September using modified naval shells.

Dirigibles at war

Historically, the first fully controllable and powered airship was French, named *La France*. First flown on 9 August 1884, the electrically powered craft made a circular flight of 5 miles (8 km). Dr Karl Wölfert designed the first successful airship to use a petrol engine, which flew in Germany in 1888. However, it was the German Count Ferdinand von Zeppelin that brought the dirigible into widespread naval use, although he is also remembered for establishing the world's first (and highly successful) commercial airline using airships, known as

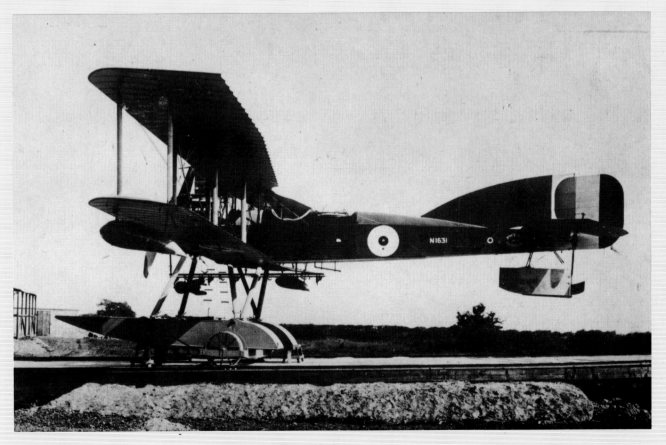

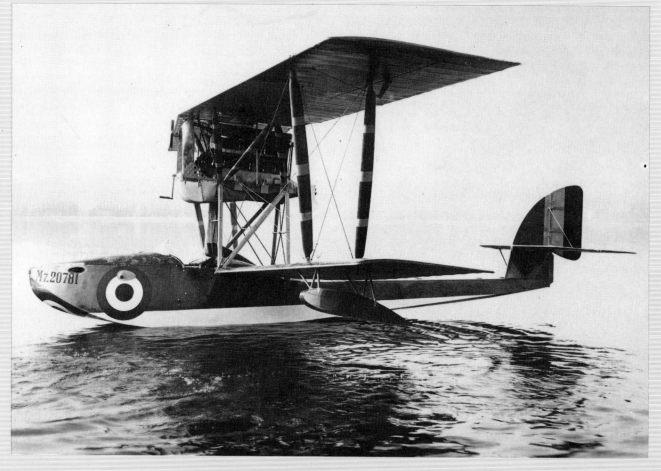

◁ ▲
*German Navy Zeppelin L59, the famous 'African' ship,
intended to carry much-needed ammunition to German
colonial forces in East Africa. It was lost on 7 April 1918,
when it burst into flames while en route to attack Royal
Navy warships at Malta.*

◁ ▼
*The 124 mph (200 km/h) single-seat Macchi M.7 flying-boat
fighter of 1917, with guns mounted in the nose.*

▲
*Early British parasite experiments centred upon the airship
R23 launching Sopwith Camel fighters, from July 1918.*

Delag and which safely carried over 34,000 passengers
between German cities from 1910 to November 1913.

Von Zeppelin's success lay in his use of rigid airships,
huge craft with complex internal structures, each with
several separate hydrogen gas bags. Britain, too, had
travelled this path with the Admiralty's Vickers RI
Mayfly, the first British rigid craft, which was 512 ft
(156 m) long but was wrecked in a handling accident on
24 September 1911 before a flight had been made. Early
Zeppelins had accidents too, and many wartime craft
were lost through action, poor weather and other reasons.

Both the German Army and Navy received Zeppelin
airships, along with similar Schütte–Lanz types, the
German Naval Airship Division operating the greatest
number with 69. It also suffered the greatest proportion
of fatalities of any branch of the German armed forces.
Naval airships provided a unique answer to the problem
of long-range strategic bombing in the early war years
and on 19 January 1915 three Navy Zeppelins launched
the first airship raid on Britain, dropping bombs on
several towns and cities, including Great Yarmouth,
where two of them released nine bombs, killing two
people. London came under attack from 31 May. It was

not until the use of Pomeroy incendiary ammunition from
guns in 1916 that high-flying airships became properly
vulnerable to attack, although one had been destroyed by
an RNAS pilot on the night of 6–7 June 1915 after he
dropped Hales bombs on to it from his Morane-Saulnier
Parasol, an epic attack given the difficulty of flying above
the airship and the constant blasts from the airship's
gunners. The last Zeppelin raid on England to cause
death or injury was on 12 April 1918. By then 51 airship
raids over the war years had killed 557 people and
injured a great many more.

However, it was not only through their bombing that
the Zeppelins proved a great nuisance; their reconnaissance flights over the North Sea repeatedly gave
away the positions of British ships and thus prevented
the German fleet from being tempted out of its bases. The
British had spent many years and a vast sum of money on
a fleet of battleships, for the express purpose of taking on
the German fleet in battle, and as long as this did not
occur Britain's main war plan was being frustrated. The
German High Seas Fleet and the British Grand Fleet did
eventually meet at the Battle of Jutland in 1916, but the
Royal Navy suffered a great deal from inadequate
reconnaissance, despite the brave work of Flight
Lieutenant F. J. Rutland, flying from the *Engadine*.

Because of the vulnerability of hydrogen gas-filled
airships to attack and bad weather, German heavy
bombers took over the bulk of strategic bombing from
mid-1917, while smaller German naval aircraft such as
the Hansa-Brandenburg seaplane fighters also carried
out much good work. Austrian Löhner flying-boats were
among many such aircraft that performed excellently in
maritime patrol and reconnaissance roles, Italian
Macchis also being small enough to be used as single- and
multi-seat fighters. It was an Austrian Löhner that
became the first aeroplane to sink a submarine, the

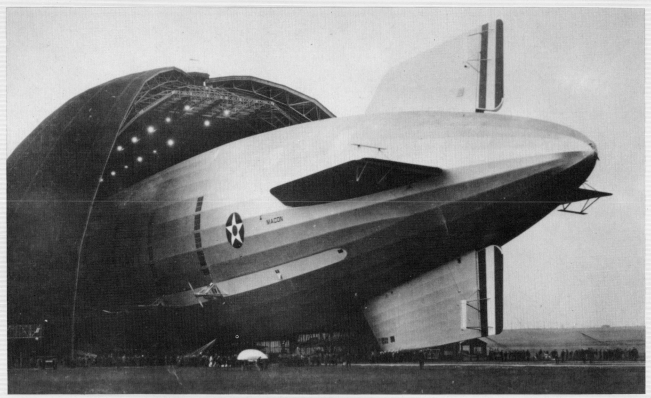

Macon emerging from its huge hangar at Akron, Ohio.

◁ ▼

A Sparrowhawk fighter engages Macon's trapeze in a mid-air retrieval.

▲

A Sparrowhawk fighter stowed in the internal hangar on board USS Akron in 1932. Note the Sparrowhawk approaching the airship in the lower centre of the photograph.

unlucky craft being the French submarine *Foucault* on 15 September 1916. British flying-boats of Felixstowe F.2A and F.3 types, operated from 1917, were derived from American Curtiss machines and conducted similar patrol and anti-submarine duties to the Curtiss America series and French FBA types and others, although the Felixstowes are also remembered for their use in early in-flight refuelling experiments.

Britain rethought her position on using rigid airships for strategic bombing only as the war closed and so this became a dead end. However, both Britain and Germany had launched fighters from airships in so-called 'parasite' experiments, a concept of airship self-protection and a way of increasing an airship's visual range that the US Navy alone put into actual operation with its Curtiss F9C-2 Sparrowhawk fighter biplane-carrying airships *Akron* and *Macon* of the 1930s. Although both eventually crashed into the sea during their consecutive periods of service with the US Navy, they were highly successful, each having an internal aircraft hangar and trapeze launching/retrieving system for the operation of up to six Sparrowhawks.

Great use was made by the combatant nations of non-rigid and semi-rigid airships, their tasks including convoy protection. After the Armistice the US Navy continued large-scale non-rigid airship operations, gaining great experience which was used again during the Second World War. The US Navy continued airship operations until the 1960s, the last craft taken into service

One of 71 Royal Navy SS Zero class non-rigid airships of 1916–18, mainly used as towed craft to spot gunfire, and here approaching the deck of HMS Furious.

being Goodyear ZPG-3Ws for airborne early warning. The envelopes were used as radomes for the huge 40 ft (12 m) radar antennae. In the 1980s the role of AEW from airships is being examined again, with Goodyear and the British company, Airship Industries, leading the way. On 27 November 1984 the US Navy staged a briefing for a battle surveillance airship system (BSAS), intended to provide an organic area surveillance system. Already by then Airship Industries had seen one of its Skyship 500s used in patrol airship concept evaluation (PACE) trials with the US Navy, US Coast Guard and NASA.

Furious and its following

The Battle of Jutland and the shortcomings experienced by the Grand Fleet were examined by various committees in 1916, and one important conclusion was that more aircraft should be sent to sea as soon as possible. In order to implement this directive, a Grand Fleet Aircraft Committee was formed and it immediately cast around for a suitably large and fast vessel which could be requisitioned and converted into an aircraft carrier.

HMS *Furious*, intended originally to be a battle cruiser but with one of its 18 in. gun turrets removed to make way for a hangar and flight-deck on the forecastle, was the first warship to be built as an aircraft carrier for landplane operations. Although subsequently modified between 1921 and 1925 to become a full flush-deck carrier with two lifts and an aircraft capacity of 33, it was converted and completed in 1917 as a carrier for six Sopwith Pups and four seaplanes. *Furious* eventually became the longest-serving carrier in the world, not being scrapped until 1949. The Pup fighter was of world class, having been used from land since 1916 and was one of the few combat planes able to fire Le Prieur rocket projectiles on anti-Zeppelin attacks. Its normal armament, however, was a Lewis or Vickers machine-gun. In an attempt to work out the techniques for landing Pups on *Furious*'s

original deck, Sqn Cdr E. H. Dunning became the test pilot. On 2 August 1917 Dunning flew his Pup to a forward and starboard position off *Furious*, which steamed at 26 knots into a 21 knot headwind. Side-slipping the fighter over the deck, he made an attempt to land while some men on deck grabbed at straps on the aircraft to bring it to rest. On 7 August Dunning attempted to repeat the landing, but stalled as he tried to overshoot the deck and his aircraft cartwheeled over the side. Meanwhile, on 11 August Flight Sub-Lieutenant Stuart Culley, RN, had taken off in his Sopwith Camel fighter from a towed barge behind HMS *Redoubt* to shoot down a Zeppelin, this method of providing some capital warships with aircraft operating ability having been the subject of experiments dating from as far back as 1912.

One outcome of an inquiry into the Dunning tragedy was that both the landing technique and the deck of *Furious* were clearly unsuited to safe operations. A second Royal Navy aircraft carrier was, therefore, not just going to need a bow deck for take-offs and a stern deck for landing, but a deck of full length and width and with no obstructions whatsoever. This second carrier, HMS *Argus*, had begun life as the Italian liner *Conte Rosso* when laid down in 1914 and had been launched in December 1917 with what was patently an unsuitable deck. Modifications were put in hand and it eventually went into service as the world's first flush-deck carrier.

During the summer and autumn of 1918, as the First World War was coming to an end, *Furious* remained as part of the Grand Fleet. It continued to fly off aircraft which then usually ditched next to it on their return instead of landing on deck, because they were considered to be dispensable and easily replaced. Her complement of aircraft now included a version of the Sopwith Camel in place of the Pups.

In June, permission was granted for a raid from *Furious* on the huge Zeppelin sheds at Tondern in Schleswig-

◄ *Airship Industries' Skyship 600 deploying a 15 ft Zodiac inflatable boat using two electric winches during search-and-rescue capability trials.*

▼ *Dunning side-slips his Pup fighter over Furious.*

Holstein, and this became the first attack ever made by ship-based landplanes (as opposed to seaplanes) against a target on land. Bad weather delayed the first attempt, but on 18 July the real attack was mounted. It was a complete surprise to the Germans and the Camels managed to destroy two airships in the largest shed. The *Furious* and its aircraft had finally achieved success.

Despite the successful Tondern bombing, British ideas

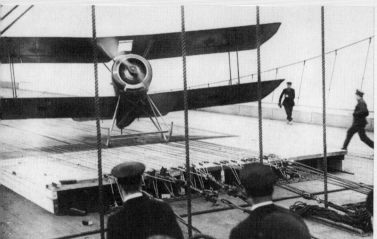

▲ HMS Furious with its second deck layout, comprising fore and aft aircraft decks and interconnecting walkways. Just to the rear of the funnel on the landing deck can be seen a structure from which ropes were fixed as a crash barrier.

◄ After Dunning's second and fatal attempt to land on Furious, the carrier was given an aft landing-on deck. In March 1918 Sqn Cmdr F. J. Rutland, RNAS, successfully landed a Pup fighter on the aft deck. As can be seen, the Pup was fitted with a skid landing gear carrying clips to engage the longitudinal fore–aft arrestor wires.

▼ Ft Sub.-Lt Stuart Culley rises from the towed lighter moments before shooting down Zeppelin L53.

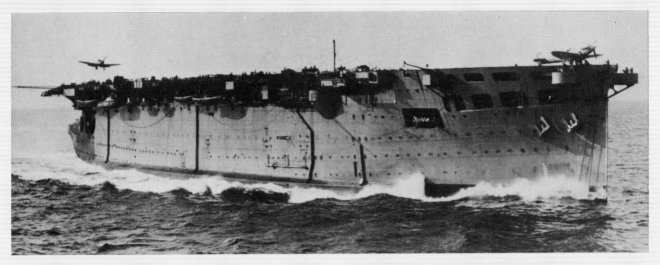

▲
HMS Argus as it appeared during its final years of service, in 1943, operating Seafires off North Africa.

▼
Britain pioneered aeroplane take-offs from gun turret platforms, first successfully attempted in mid-1917 by Ft Cmdr F. J. Rutland in a Pup fighter from the light cruiser HMS Yarmouth. It subsequently became widespread practice, and here the Imperial Japanese Navy warship Yamashiro has one of its British-built Gloster Sparrowhawk I fighters turret mounted.

changed direction and looked towards the very attractive target of the German fleet in its own bases, by using torpedoes launched from aircraft. Sopwith Cuckoos were earmarked for this task, but their promise never came to fruition because of the Armistice in November 1918.

Technical developments
The flush deck of *Argus* was ideal for pilots attempting to land on-board ship, but it was less convenient for the actual operation of the vessel, especially in terms of venting the smoke from the ship's engines and of radio communications. Another rethink led to the adoption of

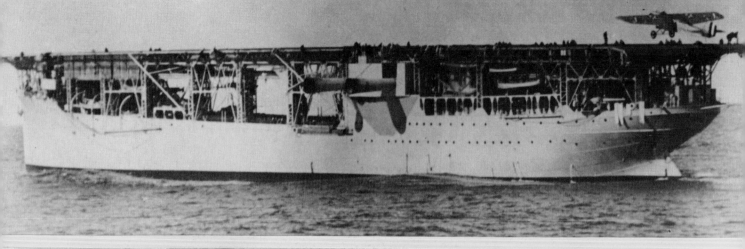

the on-deck offset island structure, development being helped by trials using *Argus* with a makeshift structure. By merging the control bridge, radio mast and engine funnels into a small structure on the starboard side of the deck (starboard was chosen as pilots tended to veer to port when in difficulty, although one of Japan's inter-war carriers, *Akagi*, had a port island structure), both the ship's and flight crews were satisfied and operations were optimized. The *Almirante Cochrane*, a battleship laid down in 1913 for Chile but taken over to become the carrier HMS *Eagle*, together with the first carrier to be laid down as such, HMS *Hermes*, were completed with starboard islands.

Another widespread method of carrying aircraft on-board capital warships was the gun turret platform, from which aircraft could be launched and later retrieved from the water. For Britain, Sopwith Pups, 1½-Strutter two-

seaters and Camels were used in this way during the war and the method became widespread afterwards with many navies.

A system of arrestor wires running along part of a carrier's deck had been adopted by the Royal Navy to bring aircraft to rest after landing, the aircraft having engaging hooks fitted to the undercarriage. This method was not entirely satisfactory and led to accidents. Aircraft such as the Fairey IIIF of 1928 service were landed unassisted, which was often considered safer, but this was because the IIIF's landing speed was only 51 mph (82 km/h). The adoption of hydraulic brakes made a difference, the Fleet Air Arm's Fairey Flycatcher fighters of 1923 service being the first to be so fitted.

The US Navy had, meanwhile, seen the conversion of the ex-collier USS *Jupiter* into its first aircraft carrier, becoming the USS *Langley* when commissioned on 20

March 1922. The first take-off from *Langley* was by a Vought VE-7SF fighter, piloted on this occasion (17 October 1922) by Lt V. C. Griffin. On 27 December the same year Japan commissioned its first carrier, *Hosho*, the first take-off being performed by a Mitsubishi 1MF1 piloted by a Briton, Capt. Jordan.

The US Navy perfected the transverse arrestor wire technique, which became and remains the standard method of halting aircraft on the deck. The US Navy followed *Langley* with the classic carriers USS *Saratoga* and *Lexington* of 1927, able to fulfil Clément Ader's dream of high speed (34 knots) and many on-board aircraft (90).

Inter-war tactical experiments
Between the two world wars, the economic depression had a devastating effect on most countries' economies,

and also drastically limited developments in the armed forces. However, Japan, the USA and Britain still managed to make advances in the operation of carrier-borne aircraft and of the carriers themselves. In particular, the USA organized a series of 'Fleet Problems' from 1928 onwards, in which their carriers were used to test the theories of their deployment. Aircraft-carrier tactics moved ahead very rapidly as a result, and it was through these wargames that the fleet commanders gained valuable experience which stood them in good stead in the Second World War.

Like the USA, Japan expended a great deal of effort on developing the concept of fast carrier strike forces; she also recognized that the USA was her chief rival, and that the US Navy had to be matched. The basis of their strategy was that the aircraft carrier had now become a major element of their fleet, rather than simply being

Fairey Swordfish II

1 Rudder structure
2 Rudder upper hinge
3 Diagonal brace
4 External bracing wires
5 Rudder hinge
6 Elevator control horn
7 Tail navigation light
8 Elevator structure
9 Fixed tab
10 Elevator balance
11 Elevator hinge
12 Starboard tailplane
13 Tailplane struts
14 Lashing-down shackle
15 Trestling foot
16 Rear wedge
17 Rudder lower hinge
18 Tailplane adjustment screw jack
19 Elevator control cable
20 External bracing wires
21 Elevator fixed tab
22 Tail fin structure
23 Bracing wire attachment
24 Aerial stub

25 Bracing wires
26 Port elevator
27 Port tailplane
28 Tailplane support struts
29 Dinghy external release cord
30 Tailwheel oleo shock absorber
31 Non-retractable Dunlop tailwheel
32 Fuselage framework
33 Arrestor hook housing
34 Control cable fairleads
35 Dorsal decking
36 Rod aerial
37 Lewis gun stowage trough
38 Aerial
39 Flexible 0.303 in. (7.7 mm) Lewis machine gun
40 Fairey high-speed flexible gun mounting
41 Type 0-3 compass mounting points
42 Aft cockpit coaming
43 Aft cockpit
44 Lewis drum magazine stowage
45 Radio installation
46 Ballast weights
47 Arrestor hook pivot
48 Fuselage lower longeron
49 Arrestor hook (part extended)
50 Aileron hinge
51 Fixed tab
52 Starboard upper aileron
53 Rear spar
54 Wing ribs
55 Starboard formation light
56 Starboard navigation light
57 Aileron connect strut
58 Interplane struts
59 Bracing wires
60 Starboard lower aileron
61 Aileron hinge
62 Aileron balance
63 Rear spar
64 Wing ribs
65 Aileron outer hinge
66 Deck-handling/lashing grips
67 Front spar
68 Interplane strut attachments
69 Wing internal diagonal bracing wires
70 Flying wires
71 Wing skinning
72 Additional support wire (fitted when underwing stores carried)

73 Wing-fold hinge
74 Inboard interplane struts
75 Stub plane end rib
76 Wing locking handle
77 Stub plane structure
78 Intake slot
79 Side window
80 Catapult spool
81 Drag struts

82 Cockpit sloping floor
83 Fixed 0.303 in. (7.7 mm) Vickers gun (deleted from some aircraft)
84 Case ejection chute
85 Access panel
86 Camera mounting bracket

87 Sliding bomb-aiming hatch
88 Zip inspection flap
89 Fuselage upper longeron
90 Centre cockpit
91 Inter-cockpit fairing
92 Upper wing aerial mast
93 Pilot's headrest
94 Pilot's seat and harness
95 Bulkhead
96 Vickers gun fairing
97 Fuel gravity tank (12.5 Imp. gal./57 litres capacity)
98 Windscreen
99 Handholds
100 Flap control handwheel and rocking head assembly

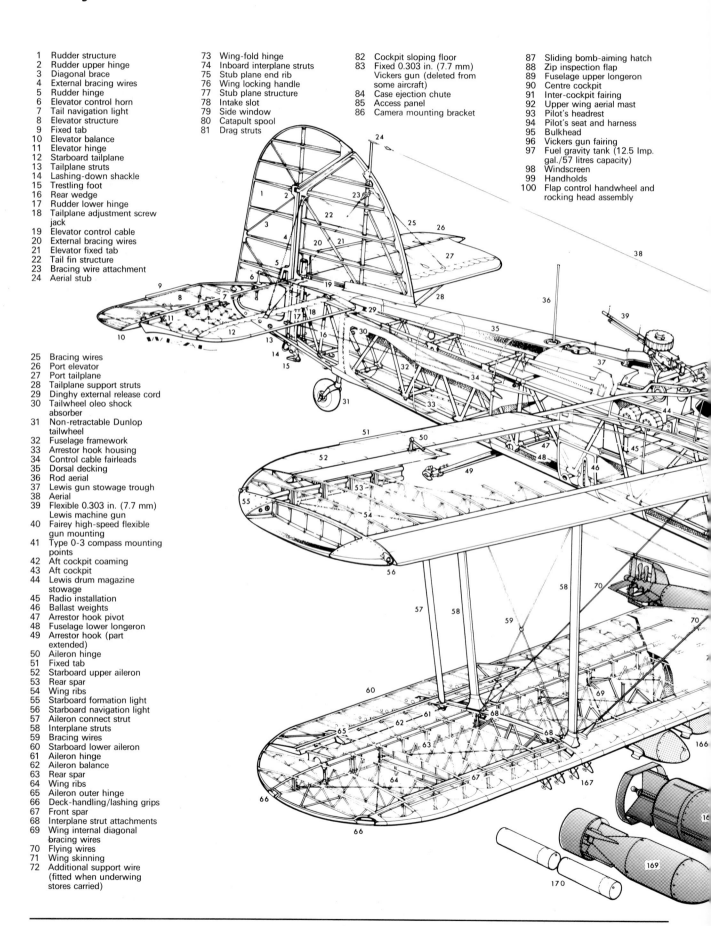

101 Wing centre-section
102 Dinghy release cord handle
103 Identification light
104 Centre-section pyramid strut attachment
105 Diagonal strengtheners
106 Dinghy inflation cylinder
107 Type C dinghy stowage well
108 Aileron control linkage
109 Trailing-edge rib sections
110 Rear spar
111 Wing rib stations
112 Aileron connect strut
113 Port upper aileron
114 Fixed tab
115 Aileron hinge
116 Port formation light
117 Wing skinning
118 Port navigation light
119 Leading-edge slot
120 Front spar
121 Nose ribs
122 Interplane struts
123 Pitot head
124 Bracing wires
125 Flying wires
126 Port lower mainplane
127 Landing lamp
128 Underwing bomb shackles
129 Underwing strengthening plate

130 Rocket-launching rails
131 Four 60 lb (27 kg) anti-shipping rocket projectiles
132 Three-bladed fixed-pitch Fairey-Reed metal propeller
133 Spinner
134 Townend ring
135 Bristol Pegasus IIIM3 (or Mk 30) radial engine
136 Cowling clips
137 Engine mounting ring
138 Engine support bearers
139 Firewall bulkhead
140 Engine controls
141 Oil tank immersion heater socket
142 Filler cap
143 Oil tank (13.75 Imp. gal./62.5 litres capacity)
144 Centre-section pyramid struts
145 External torpedo sight bars
146 Fuel filler cap

147 Main fuel tank (155 Imp. gal./705 litres capacity)
148 Vickers gun trough
149 Fuselage forward frame
150 Oil cooler
151 Fuel filter
152 Stub plane/fuselage attachment
153 Fuel feed lines
154 Dinghy immersion switch
155 Exhaust
156 Port Dunlop mainwheel
157 Jacking foot
158 1610 lb (730 kg), 18 in. (45.7 cm) torpedo

159 Access/servicing footholds
160 Torpedo forward crutch
161 Radius rod fairing
162 Undercarriage axle tube fairing
163 Undercarriage oleo leg fairing
164 Starboard mainwheel
165 Hub cover
166 Underwing bombs
167 Underwing outboard shackles
168 Depth-charge
169 250 lb (113 kg) bomb
170 Anti-shipping flares

employed in a supporting role. However, there was still strong resistance to the principle of the carrier from within the navies of these countries.

Once it became clear in the 1930s that Germany was re-arming, Britain realized she had to keep pace with the new threat from across the Channel, and she undertook an appraisal of the armed services. The Fleet Air Arm had deteriorated drastically under the management of the Air Ministry, and the worst shortcoming was the lack of modern aircraft. Also, a number of RNAS pilots had been incorporated into the Royal Air Force on its formation in April 1918, and this resulted in a majority of senior officers who considered that the battle fleets still reigned supreme. Against this, there was a return to aircraft carriers as a supporting element in the Royal Navy, but their need was hotly disputed by the RAF, who were afraid that funds for land-based aircraft would therefore be cut back. One main RAF tactic was based on high-level bombing, and this clashed directly with the requirements of naval air power. The first vessel to come from the revised approach was the *Ark Royal*, an advanced carrier but not a large one. The amount of effort put into its design and construction showed how great a change had come over the Royal Navy. Admirals were disillusioned with RAF control and finally in 1937 the government announced that it would hand back responsibility for naval aviation to the Royal Navy.

War with Germany again

At the outbreak of the Second World War in 1939, the naval situation was pretty well the same as it had been in the First World War, with Britain attempting to prevent the main German fleet from escaping into the North Atlantic. The Admiralty decided to take the fight to the U-Boats by forming 'hunting groups' made up of a single aircraft carrier and four destroyers, so that the carriers' aircraft could protect the supply convoys from U-Boat attack in the Western Approaches to Britain. This policy was shown to be extremely risky when the *Ark Royal* narrowly escaped being torpedoed by a U-Boat on 14 September 1939, but the *Courageous* was not so lucky three days later; it was sent to the bottom incurring a large loss of life, and the remaining carriers were withdrawn immediately. Contrary to popular opinion, the loss of the carrier was far graver than that of the battleship *Royal Oak* in October.

Both the *Furious* and *Glorious* (later joined by *Ark Royal*) were involved in the Norwegian campaign in 1940, which by the June of that year was collapsing. The surviving RAF fighters flew from the Norwegian mainland on to *Glorious* to prevent them being destroyed on the ground; amazingly, every one landed safely despite the fact that none of the pilots had ever flown on to a carrier before. But *Glorious* was extremely vulnerable in those waters, and she was hit by shells from the German battlecruisers *Scharnhorst* and *Gneisenau*, in spite of the efforts of her two escorts. The loss of a second carrier was bad enough, but the experienced RAF aircrews and valuable planes went down with her too, and few men were saved.

Once the Norwegian campaign had failed, the *Ark Royal* headed for the Mediterranean to form the core of a force intended to prevent the Italian fleet breaking out into the Atlantic, and to patrol the areas left vacant by the French (who were under German control by this time). *Ark Royal*'s Swordfish attempted to torpedo some of the French fleet at Mers-el-Kebir, in order to prevent them falling into German control, and then it helped to provide air cover during the Dakar operation, despite its aircraft being outclassed. In between these two operations, *Ark Royal* demonstrated the valuable flexibility of the aircraft carrier by helping in the operation to reinforce Malta; it gave air cover to convoys and some of its Swordfish attacked targets on the island of Sardinia, trying to provoke the Italians into defending 'their sea'.

Admiral Cunningham, commander of the Mediterranean Fleet, was responsible for causing the Italians many problems after this, and he was given permission to use an aircraft carrier for offensive action. With minor losses, his small force destroyed or damaged 40 Italian aircraft in the four months to January 1941. In addition, the carriers *Eagle* and *Illustrious* roamed the Central Mediterranean, their aircraft sinking shipping, striking airfields and mining harbours.

Encouraged by early successes, Cunningham resurrected the plan conceived in 1918, a strike against a fleet in port. The target was the main Italian base at Taranto, and it finally took place on 11 November 1940 from *Illustrious*. Although the base was not taken by surprise, the raid was well co-ordinated and inflicted considerable damage. The greatest damage, however, was to Italian morale, as half their fleet had been put out of action. For the loss of only two Swordfish, the *Littorio* was put out of action for over six months, the *Diulio* for eight months, and the *Conte di Cavour* was found to be irreparable.

In their reinforcement of Malta, the carriers were aided by the fact that they were in possession of radar and the fighter direction techniques first employed successfully in the Battle of Britain. It was the radar techniques which helped to prepare the ground for US success against the Japanese in the Pacific.

Pearl Harbor

The effectiveness of carrier raids had been tested by the Japanese in 1940, and they studied the results of the Taranto raid with great interest; the British success provided Admiral Yamamoto with the encouragement needed to undertake an attack on US forces in the Pacific Ocean, in particular at the forward base of the Pacific Fleet at Pearl Harbor, Hawaii. The surprise strike against Pearl Harbor was devastating, and was the first really large attack ever mounted by a carrier strike force. Unfortunately for the Japanese, the carriers *Saratoga*, *Lexington* and *Enterprise* were at sea at the time, and so escaped. Also, the bombers neglected to destroy the vital oil tanks and repair shops, which would have been a severe blow to the US forces in the Pacific.

The disaster at Pearl Harbor was a landmark in the use of carriers by the Americans. The destruction of the battle fleet meant that deployment of the aircraft carriers was the only way of taking the fight to the Japanese, and from then on the US offensive was based on the carrier task force. However, for six months the Japanese war machine rolled across the Pacific territories, carrier aircraft even attacking Darwin in northern Australia in February 1942.

◄ Imperial Japanese Navy Aichi D3As were the first aircraft to attack Pearl Harbor on 7 December 1941.

▼ USS West Virginia and USS Tennessee engulfed in flames at Pearl Harbor.

Mitsubishi A6M2 'Zero'

1 Tail navigation light
2 Tail cone
3 Tail-fin fixed section
4 Rudder lower brace
5 Rudder tab (ground adjustable)
6 Fabric-covered rudder
7 Rudder hinge
8 Rudder post
9 Rudder upper hinge
10 Rudder control horn (welded to torque tube)
11 Aerial attachment
12 Tail-fin leading edge
13 Forward spar
14 Tail-fin structure
15 Tail-fin nose ribs
16 Port elevator
17 Port tailplane
18 Piano-hinge join
19 Fuselage dorsal skinning
20 Control turnbuckles
21 Arrestor hook release/retract steel cable runs
22 Fuselage frame/tailplane centre-brace
23 Tailplane attachments
24 Elevator cables
25 Elevator control horns/torque tube
26 Rudder control horns
27 Tailwheel combined retraction/shock strut
28 Elevator trim tab
29 Tailwheel leg fairing
30 Castored tailwheel
31 Elevator frame (fabric-covered)
32 Elevator outer hinge
33 Tailplane structure
34 Forward spar
35 Elevator trim tab control rod (chain-driven)
36 Fuselage flotation bag rear wall
37 Arrestor hook (extended)
38 Arrestor-hook pivot mounting
39 Elevator trim tab cable guide
40 Fuselage skinning
41 Fuselage frame stations
42 Arrestor-hook position indicator cable (duralumin tube)
43 Rudder cables
44 Elevator cables
45 Trim tab cable runs
46 Arrestor-hook pulley guide
47 Fuselage stringers
48 Fuselage flotation bag front wall
49 Fuselage construction join
50 Wing-root fillet formers
51 Compressed-air cylinder (wing gun charging)
52 Transformer
53 'Ku'-type radio receiver
54 Oxygen cylinder (starboard); CO_2 fire-extinguisher cylinder (port)
55 Battery
56 Radio tray support
57 Radio transmitter
58 Canopy/fuselage fairing
59 Aerial mast support/lead-in
60 Aerial
61 Aerial mast (forward-raked)
62 Canopy aft fixed section
63 Aluminium and plywood canopy frame
64 Crash bulkhead/headrest support
65 'Ku'-type D/F frame antenna mounting (late models)
66 Canopy track
67 Turnover truss
68 Pilot's seat support frame
69 Starboard elevator control bell-crank
70 Aileron control push-pull rod

71 Wing rear spar/fuselage attachment
72 Fuselage aft main double frame
73 Aileron linkage
74 Landing-gear selector lever
75 Flap selector lever
76 Seat adjustment lever
77 Pilot's seat
78 Cockpit canopy rail
79 Seat support rail
80 Elevator tab trim handwheel
81 Fuel gauge controls
82 Throttle quadrant
83 Reflector gunsight mounting (offset to starboard)
84 Sliding canopy
85 Plexiglas panels
86 Canopy lock/release
87 Windscreen
88 Fuselage starboard 0.303 in. (7.7 mm) machine gun
89 Control column
90 Radio control box
91 Radio tuner

92 Elevator control linkage
93 Rudder pedal bar assembly
94 Cockpit underfloor fuel
95 Wing front spar/fuselage attachment
96 Fuselage forward main double frame
97 Ammunition magazine
98 Ammunition feed
99 Blast tube
100 Cooling louvres
101 Fuselage fuel tank, capacity 34 Imp. gal. (155 litres)
102 Firewall bulkhead
103 Engine bearer lower attachment
104 Engine bearer upper attachment
105 Oil tank, capacity 12.7 Imp. gal. (58 litres)

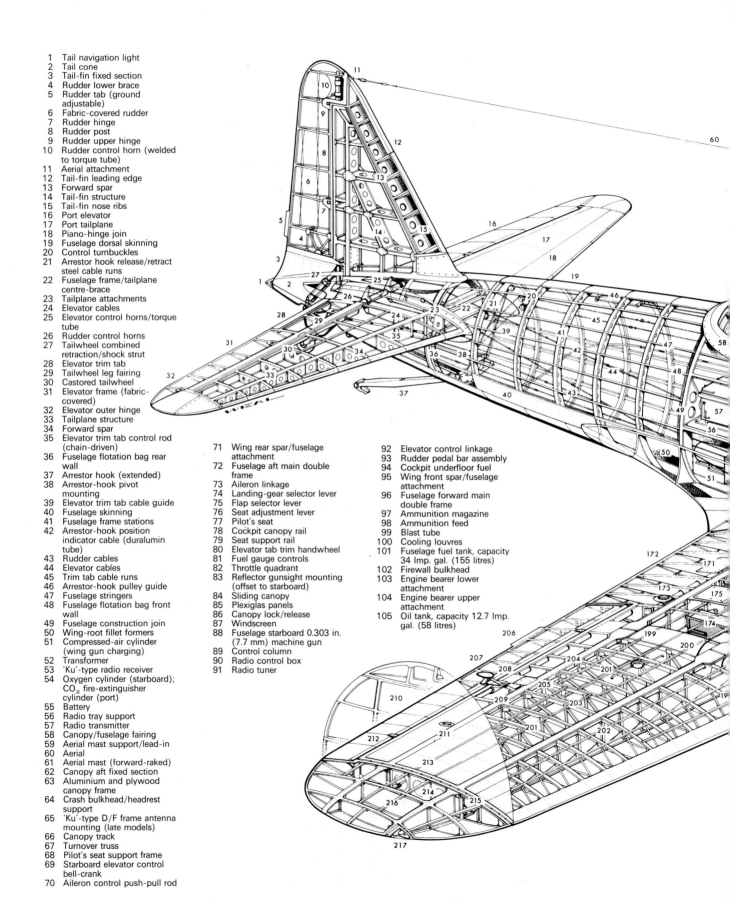

And so it begins . . .

106 Bearer support struts
107 Cowling gill adjustment control
108 Machine gun muzzle trough
109 Barrel fairing
110 Oil filler cap
111 Fuselage fuel tank filler cap
112 Port flap profile
113 Port fuselage machine gun
114 Port wing gun access panels
115 Port inner wing identification light
116 Port wing flotation bag inner wall
117 Wing spar joins
118 Aileron control rods
119 Port aileron (fabric-covered)
120 Aileron tab (ground adjustable)
121 Aileron external counter-balance
122 Control linkage
123 Wing skinning
124 Port outer wing identification light
125 Port navigation light lead conduit
126 Wingtip hinge
127 Wing end rib
128 Port wing flotation bag outer wall
129 Wing-tip structure
130 Port wing-tip (folded)
131 Port navigation light
132 Port wing-tip hinge release catch
133 Pitot head
134 Wing leading-edge skinning
135 Wing front spar
136 Port wing gun muzzle
137 Port undercarriage visual indicator
138 Undercarriage hydraulics access
139 Nacelle gun troughs
140 Cowling gills
141 Fuselage gun synchronization cable
142 Bearer support strut assembly

159 Oil cooler
160 Wing-root fasteners
161 Starboard mainwheel well
162 Front auxiliary spar cut-outs
163 Auxiliary fuel tank, capacity 74 Imp. gal. (337 litres)
164 Cockpit air intake
165 Intake trunking
166 Front main spar
167 Starboard wing fuel tank, capacity 43 Imp. gal. (195 litres)
168 Fuel filler cap
169 Rear main spar
170 Flap actuating cylinder
171 Access cover
172 Starboard flap structure
173 Starboard inner wing identification light
174 Starboard wing 20 mm machine gun

189 Mainwheel door fairing
190 Axle hub
191 Access plate
192 Hinge
193 Leg fairing attachments
194 Brake line
195 Leg fairing
196 Leg fairing upper flap
197 Wing gun barrel support collar

143 Carburettor
144 Exhaust manifold
145 Cowling panel fastener clips
146 925 h.p. Nakajima Sakae 12 radial engine
147 Cowling inner ring profile
148 Cowling nose ring
149 Three-bladed propeller
150 Spinner
151 Propeller gears
152 Hub
153 Carburettor intake
154 Port mainwheel
155 Oil cooler intake
156 Exhaust outlet
157 Starboard mainwheel inner door fairing
158 Engine bearer support brace

175 Access panels
176 Ammunition magazine (underwing loading)
177 Landing-gear hydraulic retraction jack
178 Hydraulic lines
179 Starboard undercarriage visual indicator
180 Landing-gear pivot axis
181 Undercarriage/spar mounting
182 Starboard wing gun muzzle
183 Starboard undercarriage leg
184 Oleo travel
185 Welded steel wheel fork
186 Wheel uplock latch
187 Starboard mainwheel
188 Wheel door fairing ball-and-swivel closure

198 Wing nose ribs
199 Wing spar joins
200 Cartridge ejection chute
201 Wing outer structure
202 Front spar outer section
203 Inter-spar ribs
204 Rear spar outer section
205 Aileron control access
206 Aileron (ground adjustable)
207 Starboard aileron frame
208 Aileron external counter-balance
209 Control linkage
210 Starboard wing-tip (folded)
211 Starboard outer wing identification light
212 Aileron outer hinge
213 Starboard wing flotation bag outer wall
214 Wing end rib
215 Starboard wing-tip hinge release catch
216 Wing-tip structure
217 Starboard navigation light

▲

Douglas TBD-1 Devastator, the first US Navy aircraft with hydraulically folding wings.

▲▶

USS Yorktown *sinks after being hit by bombs and torpedoes in two separate Japanese attacks during the Battle of Midway.*

▶

Grumman Avengers showing their dorsal gun turrets and ventral gun positions. Forward-firing guns completed the fixed armament, which complemented an internal bomb load.

When the Americans discovered (by breaking the Japanese naval code) that a Japanese invasion of New Guinea and the Solomon Islands was imminent, they tried to prevent this happening, and the result was the 'Battle of the Coral Sea' in May 1942. This was the first time in history that opposing fleets had not sighted each other during an engagement: it was exclusively a carrier action. In the battle, *Lexington* was lost, along with 69 aircraft, while the Japanese lost *Shoho* and 85 aircraft, and *Shokaku* was so badly damaged that it was not available for the forthcoming Battle of Midway.

One of the aircraft used by the US Navy during Coral Sea was the Douglas TBD Devastator, delivered for service from 1937 and remembered also as the first US naval aircraft with hydraulically folding wings. Equally importantly, the first single-engined carrier plane with a power-operated gun turret was the superb Grumman TBF Avenger torpedo bomber, while Grumman F4F Wildcat and F6F Hellcat fighters were highly significant to both US and British carrier forces, among other types.

A second engagement soon after Coral Sea was the

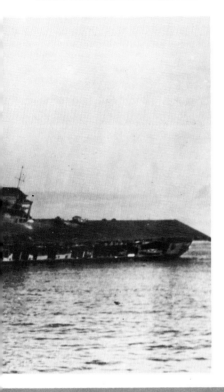

▲

One solution to the problem of providing fighter protection to convoys in the Atlantic and to deal with the menace of long-range maritime reconnaissance bombers, such as the German Focke-Wulf Fw 200 Condor, was the British CAM (catapult armed merchant) ship. Here a Sea Hurricane Mk Ia of the Merchant Service Fighter unit blasts from a CAM ship, its pilot having to bale out and parachute back to the convoy after making the interception or make for any nearby land.

Battle of Midway in June, which turned out to be the greatest US Navy success of the Second World War. In this battle, the Japanese lost all four participating carriers, for the loss of *Yorktown* to the US Navy. The Japanese also lost a large proportion of their trained and combat-experienced pilots, and Yamamoto called off the invasions. The tide had turned in the Pacific War.

The sweep across the Pacific Ocean by Japanese forces had taught the US Navy that a carrier force could overwhelm enemy air power in a particular area, and prevent reinforcements by maintaining control of the air. To this end the Navy organized carrier task forces comprising one or more task groups. These were made up of a few carriers of varying sizes, and an escort of battleships, cruisers and destroyers. The benefit of the task group was that it increased defensive power by providing a greater concentration of anti-aircraft fire, it provided a better measure of protection against submarines, and allowed a smaller number of fighters to supply effective air defence than if the carriers had operated singly. Radar was the scientific advance which made the task group possible, because it gave early warning of air attacks and allowed the component ships of the force to operate close together even in bad visibility.

Curtiss SB2C-4 Helldiver

1 Curtiss Electric four-bladed constant-speed propeller
2 Spinner
3 Propeller hub mechanism
4 Spinner backplate
5 Propeller reduction gearbox
6 Carburettor intake
7 Intake ducting
8 Warm air filters
9 Engine cowling ring
10 Oil cooler intake
11 Engine cowlings
12 Wright R-2600-20 Cyclone 14 radial engine
13 Cooling air exit louvres
14 Exhaust collector
15 Exhaust pipe fairing
16 Oil cooler

60 Fuel tank filler cap
61 Fuselage fuel tank (110 US gal./416 litres capacity)
62 Fuselage main longeron
63 Handhold
64 Fuselage frame and stringer construction
65 Autopilot controls
66 Sliding canopy rail
67 Aerial lead-in
68 Radio equipment bay
69 Life raft stowage
70 AT-4/ARN-1 transmitter aerial
71 Gunner's forward sliding canopy cover

17 Engine accessories
18 Hydraulic pressure accumulator
19 Boarding step
20 Cabin combustion heater
21 Engine oil tank (25 US gal./94.6 litres capacity)
22 Engine bearer struts
23 Hydraulic fluid tank
24 Fireproof engine compartment bulkhead
25 Aerial mast
26 Starboard wing-fold hinges
27 Wing-fold hydraulic jack
28 Gun camera
29 Rocket projectiles (4.5 in./ 11.43 cm)
30 Starboard leading-edge slat (open)
31 Slat roller tracks
32 Slat operating cables
33 Starboard navigation light
34 Formation light
35 Starboard aileron
36 Aileron aluminium top skins
37 Aileron control mechanism
38 Starboard dive brake (open position)
39 Windshield
40 Bullet-proof internal windscreen
41 Reflector gunsight
42 Instrument panel shroud
43 Cockpit coaming
44 De-icing fluid tank
45 Instrument panel
46 Pilot's pull-out chart board
47 Rudder pedals
48 Control column
49 Cockpit floor level
50 Engine throttle controls
51 Pilot's seat
52 Oxygen bottle
53 Safety harness
54 Armoured seat back
55 Headrest
56 Pilot's sliding cockpit canopy cover
57 Jury strut
58 Wing folded position
59 Fixed bridge-section between cockpits

72 Gun mounting ring
73 Gunner's seat
74 Footrests
75 Ammunition boxes
76 Armour plate
77 Wind deflector
78 Twin 0.3 in. (7.62 mm) machine guns

79 Retractable turtle decking
80 Gun rest mounting
81 Folding side panels
82 Upper formation light
83 Fin root fillet
84 Starboard tailplane
85 Deck handling handhold
86 Fabric-covered elevator

87 Remote compass transmitter
88 Tail fin construction
89 Aerial cable
90 Sternpost
91 Rudder construction
92 Fabric skin covering
93 Trim tab
94 Balance tab

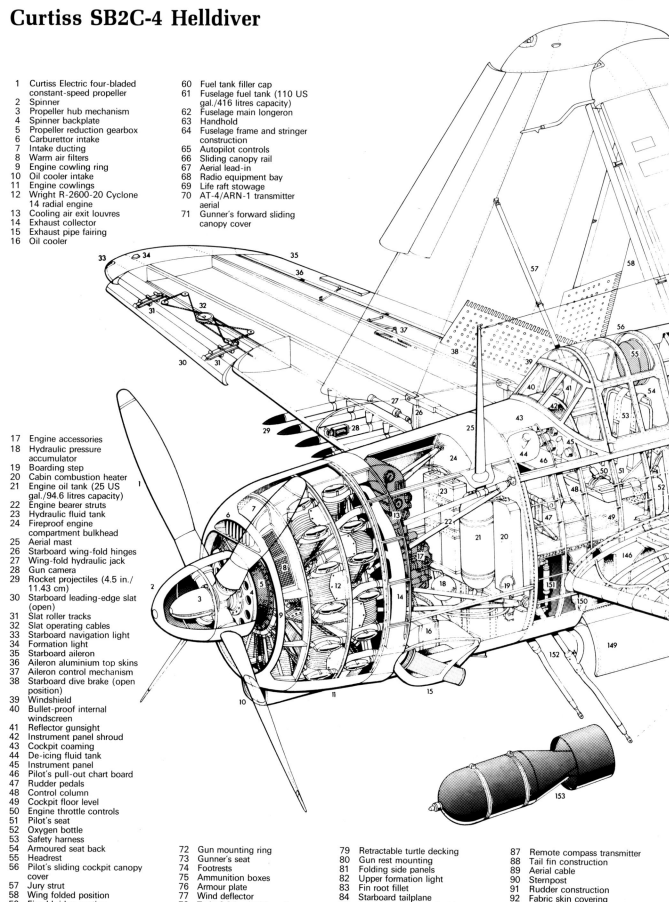

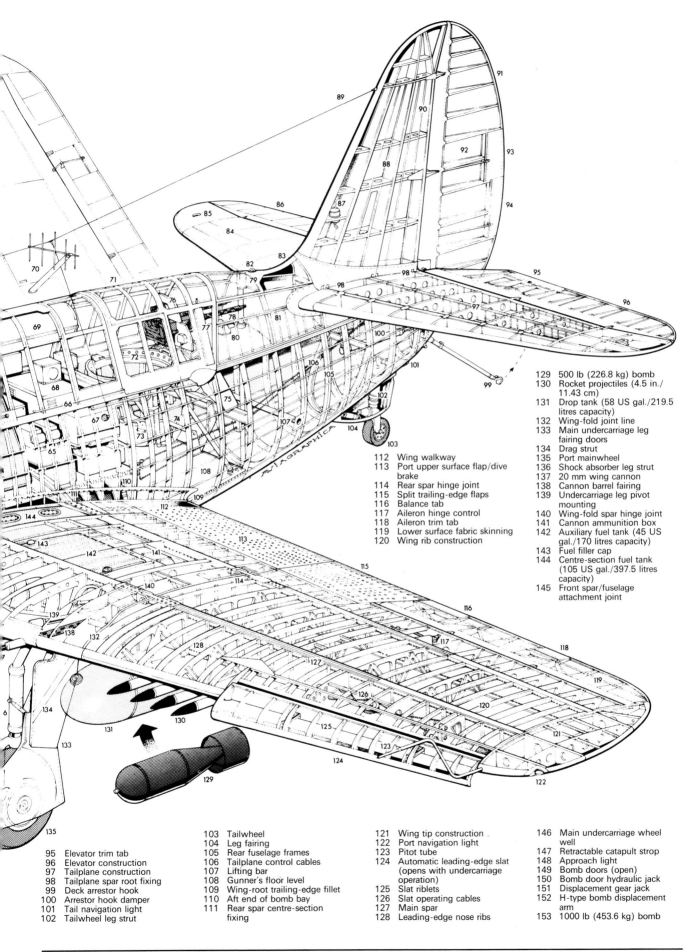

129 500 lb (226.8 kg) bomb
130 Rocket projectiles (4.5 in./
     11.43 cm)
131 Drop tank (58 US gal./219.5
     litres capacity)
132 Wing-fold joint line
133 Main undercarriage leg
     fairing doors
134 Drag strut
135 Port mainwheel
136 Shock absorber leg strut
137 20 mm wing cannon
138 Cannon barrel fairing
139 Undercarriage leg pivot
     mounting
140 Wing-fold spar hinge joint
141 Cannon ammunition box
142 Auxiliary fuel tank (45 US
     gal./170 litres capacity)
143 Fuel filler cap
144 Centre-section fuel tank
     (105 US gal./397.5 litres
     capacity)
145 Front spar/fuselage
     attachment joint

112 Wing walkway
113 Port upper surface flap/dive
     brake
114 Rear spar hinge joint
115 Split trailing-edge flaps
116 Balance tab
117 Aileron hinge control
118 Aileron trim tab
119 Lower surface fabric skinning
120 Wing rib construction

95   Elevator trim tab
96   Elevator construction
97   Tailplane construction
98   Tailplane spar root fixing
99   Deck arrestor hook
100 Arrestor hook damper
101 Tail navigation light
102 Tailwheel leg strut

103 Tailwheel
104 Leg fairing
105 Rear fuselage frames
106 Tailplane control cables
107 Lifting bar
108 Gunner's floor level
109 Wing-root trailing-edge fillet
110 Aft end of bomb bay
111 Rear spar centre-section
     fixing

121 Wing tip construction
122 Port navigation light
123 Pitot tube
124 Automatic leading-edge slat
     (opens with undercarriage
     operation)
125 Slat riblets
126 Slat operating cables
127 Main spar
128 Leading-edge nose ribs

146 Main undercarriage wheel
     well
147 Retractable catapult strop
148 Approach light
149 Bomb doors (open)
150 Bomb door hydraulic jack
151 Displacement gear jack
152 H-type bomb displacement
     arm
153 1000 lb (453.6 kg) bomb

Preceding pages
A carrier-based dive-bomber, the Helldiver played an important part in the Pacific island campaigns against the Japanese from late 1943.

▲
Lt Col James Doolittle takes the Pacific War to the Japanese homeland, leading a force of 16 Mitchell bombers on a one-way raid to attack Tokyo, Yokohama, Yokosuka, Nagoya and Kobe.

The development of replenishment at sea meant that carriers could remain at sea for as long as two months.

The USA began its main Pacific offensive at the end of 1943 and within a year it had pretty well destroyed the Japanese navy. The US Navy had a huge force, later helped by British carriers, and the fleet that attacked Tokyo in February 1945 was equipped with 1200 aircraft. As a morale booster, and as a psychological blow to the Japanese, a force of 16 Mitchell bombers had made a strike on Tokyo and other cities as early as 18 April 1942. Lt Col James Doolittle flew with his colleagues from USS *Hornet* for this one-way mission – for those that reached it, China was the landing ground.

A clear indication of the effectiveness of US carrier-borne aircraft in the Pacific during the Second World War is given by the statistic that they destroyed 15,000 enemy aircraft, 161 Japanese navy ships and more than 60 German U-Boats. In addition, nearly 1,400,000 tons of merchant shipping was sunk, a figure beaten only by US submarines. American carrier losses in order to achieve this total were 11, together with over 450 aircraft. It is extremely interesting that, once the task force system had been introduced in 1943, neither the Royal Navy nor the US Navy lost a main fleet carrier, despite the efforts of Japanese kamikaze units late in the war.

▲
*Captured by the cameraman a
moment before it ends its attack, a
Japanese Zero kamikaze plane
dives for USS Missourri, 28 April
1945.*

▶

*The badly damaged deck of the
auxiliary aircraft carrier USS
Sangamon (ACV-26) following a
kamikaze attack in 1945.*

◁

*A Fleet Air Arm Swordfish
prepares to take off from the
snow-covered deck of HMS Fencer
during convoy escort duty to
Russia.*

◀

CVS-36 USS Antietam, the first
aircraft carrier built with the
British-developed angled deck.

▼◀

A post-war Royal Navy carrier
showing the mirror light landing
system on the opposite side of the
deck from the superstructure.

Post-war developments

By the end of the Second World War, on-deck fighters
possessed high landing approach speeds, and the advent
of turbojet-powered warplanes meant that future naval
warplanes would be both faster and heavier, leading to
today's 'meteors of metal' like the Tomcat, landing at well
over 30 tons weight. Never again would aircraft such as
the 138 mph (222 km/h) Fairey Swordfish be found on
deck, even though this pedestrian of the sky performed
excellent work throughout the war.

Although the US Navy was eventually to dominate
post-1945 organic air power at sea, Britain led post-war
developments with four major innovations: the angled
deck, which allowed aircraft taking off to be readied to
the port side of the deck while freeing the remainder for
landing operations and making it much safer if incoming
aircraft missed the arrestor wires; the mirror light
landing system which enabled an incoming pilot to judge
his approach precisely by viewing a pattern of lights in a
mirror to one side of the deck; the steam catapult which
superseded the hydraulic catapult and was powered by
the ship's own main boilers; and the so-called ski-jump,
as detailed in Chapter 6. The Royal Navy has never
introduced nuclear power to its carriers, which was first
used on board the carrier USS Enterprise in 1961, having
been commissioned for the first time on the cruiser USS
Long Beach.

Post-war jet warplanes rapidly grew in size and
weight, the Douglas A3D Skywarrior becoming the
heaviest aircraft to serve on a carrier at a weight of 82,000
lb (37,195 kg), although in 1963 USS Forrestal was used
in a series of experiments in which a Lockheed KC-130F
Hercules tanker performed 21 unassisted take-offs and
landings at a gross weight of up to 120,000 lb (54,430 kg).
The largest and fastest attack bomber to serve on a
carrier was the Mach 2 North American A-5 Vigilante,
which later served in a reconnaissance role.

And so it begins . . .

▶ *Supermarine Scimitar strike-fighter prototype about to be launched by steam catapult. The aircraft's new blown flaps helped to reduce catapult launch speed.*

▼ *Assisted launch of an A3D-2 Skywarrior of Navy Squadron VAH-10.*

The development of the atomic bomb during the Second World War, the high performance and long range of post-war bombers and the vulnerability of airfields and carrier decks, all led to new concepts being tested to make air forces survivable. The latter problem had been adequately demonstrated by the 2257 Japanese suicide attacks on Allied shipping in the Pacific during the war. These destroyed or damaged 322 US ships alone from the formation of the *Shimpu* Special Attack Corps in October 1944 (with Zero fighter conversions) to the last kamikaze attack on 15 August 1945.

One answer was an attempt to develop flying-boat fighters, such as the US Convair Sea Dart of 1953, but this concept was later abandoned. Helicopters had first come to the deck on-board wartime German warships, while the German Navy had also deployed a rotorkite for observation from surfaced submarines from 1942. But it

▲
On 3 July 1950 US Navy Grumman F9F-2 Panther fighters operating from USS Valley Forge became the first American jets to fly into action, against North Korean forces.
▷ ▲
The Convair Sea Dart experimental seaplane fighter during its take-off transition from taxiing on its hull to aquaplane speed on its retractable skis.
▷ ▶
A Vigilante assigned to USS Forrestal.

was early trials on-board US ships that proved the value of helicopters, beginning when the US Army Air Force (USAAF) tested a Sikorsky R-4 on-board the tanker *Bunker Hill*, with the War Shipping Administration and US Coast Guard in May 1943.

Tactics

When the Second World War ended in 1945, both the Royal Navy and the US Navy severely reduced their carrier fleets to a level commensurate with peacetime requirements. However, for the USA, various Cold War crises and the Korean and Vietnam wars precipitated a strengthening of naval air power, leading to the 1960s' decision to maintain a minimum of 15 attack carriers.

In the Korean War, one British and one US carrier were used for land strikes in mid-1950, but their main role was to support land fighting and disrupt lines of communication from North Korea. This contribution by the US Navy was continued during the Vietnam War of later years.

The above is just the tip of the iceberg in naval air power history. The men and machines, the inventions, judgements and wars have all played a part in making naval air power what it is today, probably the most flexible aspect of the world's armed forces.

▲

On 21 July 1946 the prototype McDonnell FH-1 Phantom, piloted by Lt Cmdr James J. Davidson, landed on board USS Franklin D. Roosevelt to become the first pure turbojet-powered aircraft to operate from a carrier deck.

▷ ▲

The only rotorcraft ever to serve operationally from a submarine was the German Focke-Achgelis Fa 330 Bachstelze towed and unpowered observation rotor-kite, which went into service in 1942 but was unpopular.

▷ ▼

Vietnam saw the first widescale use of helicopters in armed form, US forces following the earlier French lead in this field. Here a Marine Corps Iroquois attempts to suppress ground fire with machine-gun fire and rockets.

Chapter 2
Carrier battle groups
– key to Atlantic security

In the years following the Second World War, many naval planners believed that the era of war at sea involving heavy surface combatants such as aircraft carriers and battle cruisers was over – the submarine now ruled the waves. A number of British politicians, for example, faced with strained resources for defence in the 1960s and 1970s, took this view. Plans to build a new large fleet aircraft carrier for the Royal Navy were scrapped; the UK's once-powerful fleet of 'flat-tops' was cut almost to the point of the Fleet Air Arm (FAA) becoming extinct, and 'organic air' at sea looked as if it would be represented solely by anti-submarine helicopters deployed on frigates and destroyers. In the event, the situation was saved by the development of the Sea Harrier, a version of the RAF's V/STOL strike aircraft, coupled with a new breed of warship termed the 'through-deck cruiser', which became the *Invincible* class of aircraft carrier. The Falklands war between Britain and Argentina in 1982 once again proved the value of gaining sea control using carrier air power, in this case in order to isolate Argentine forces on the islands and to provide potent air attack against ground targets.

The Soviet Union has never seemed in doubt about the value of the heavy surface combatant, although until the 1970s it appeared little interested in the conventional aircraft carrier as a means of power projection. Admiral of the Fleet, Sergei Gorskhov, so-called Father of the modern Soviet Navy, wrote: 'battle at sea always was and remains the main means of solving tactical tasks'. The current Soviet naval construction programme echoes this theme. The fourth *Kiev* class carrier, *Kharkov*, is shortly to join the fleet after completing sea trials; the first of up to eight nuclear-powered strike carriers is being built, as is the third *Kirov* class nuclear-powered battle cruiser.

The US Navy, with the support of a powerful lobby in Congress, has fought off attempts by other sections of the armed forces and by some politicians to cut funding for its carrier programme, despite the vast cost, although during the late 1970s the completion of CVN 71 (*Theodore Roosevelt*) did seem in doubt. As part of the President Reagan administration's effort to halt the decline of the US Navy in the years following the Vietnam war, a force of nearly 600 ships has been achieved, with plans for 15 'urgently required' separately-deployable carrier battle groups (CBGs) well on their way to fruition. There has been a change of thinking among US strategists too. A revised stance is being discussed to meet changing situations, with naval power being deployed to the Norwegian Sea, spearheaded by CBGs.

In the 1980s, sea control is all important. Despite the might of the US Navy, and the combined strength of the other NATO navies, the vital North Atlantic can no longer be safely considered as a NATO 'pond'. Naval planners classify command of the sea on a sliding scale of one to five, with one representing absolute control and friendly forces operating in complete freedom, and five being the complete opposite with enemy ships and aircraft holding total sway. At present, the balance of forces between NATO and the Warsaw Pact in this sea-power equation

Preceding pages

For medium attack, the A-6E Intruder is found on board US carriers.

British Aerospace AV-8 Harrier, AV-8S Matador and Sea Harrier (UK)

Developed from a series of experimental V/STOL fixed-wing aircraft known as P.1127s and Kestrels, the production Harrier eventually went into RAF service from 1969 as a close-support and tactical reconnaissance aircraft. The 'heart' of the Harrier is its Pegasus vectored-thrust turbofan engine, which exhausts through four nozzles that can be rotated through 98.5° to achieve vertical and forward/rearward horizontal flight or short take-offs and landings. Although RAF Harriers, with nose-mounted laser rangers and marked target seekers in 'thimble' nose extensions, were flown from Royal Navy ships during the Falklands campaign (notably from HMS *Hermes*), these are not generally used in the accepted naval context. However, the US Marine Corps previously received 102 single-seat Harriers and eight two-seaters without laser rangers and marked target seekers, in US service being known as AV-8As and TAV-8As respectively. Of these, 47 were later modified to AV-8C standard by the addition of warning radar, flare/chaff dispensers, lift improvement devices and much else to bring them more in line with newly ordered Harrier IIs. In 1985, three USMC squadrons had 45 AV-8A/Cs in active use, plus the trainers. Meanwhile, the Spanish Navy received 11 similar AV-8S Matadors and two two-seat TAV-8Ss; nine of the Matadors are currently in active use with 8a Escuadrilla. Five single-seaters and the two-seaters are normally on board the aircraft carrier *Dédalo*. From the Harrier, British Aerospace developed the Sea Harrier for the Royal Navy, to suit the forthcoming *Invincible* class aircraft carriers. Sea Harrier FRS. 1s joined the Royal Navy in 1979 and in 1982 fought with great distinction during the Falklands campaign (see Chapter 6). Unlike the Pegasus Mk 103 engines of RAF Harriers and Pegasus 803s (F402-RR-402/402As) in USMC Harriers, the Sea Harrier uses the Pegasus Mk 104 and has no magnesium components, the cockpit is raised, the avionics are revised (including use of Ferranti Blue Fox multi-mode radar in the nose), and it has the standard provision for Sidewinder air-to-air missiles (as do USMC Harriers and Spanish Matadors). The Royal Navy currently operates 29 Sea Harrier FRS.1s plus a few T.Mk 4N non-navalized two-seat trainers, with more Sea Harriers on order (see Chapter 6). The Indian Navy has received some of the 18 FRS.51 single-seat Sea Harriers ordered and two two-seaters.

Data: Sea Harrier FRS.1
Number in crew: One
Engine: One 21,500 lb (9752 kg) thrust Rolls-Royce
   Pegasus Mk104 vectored-thrust turbofan
Length: 47 ft 7 in. (14.50 m)
Wing span: 25 ft 3 in. (7.70 m)
Gross weight: 26,200 lb (11,880 kg)
Maximum speed: above 736 mph (1185 km/h)
Combat radius: 288 miles (463 km)
Weapons: Four Sidewinder air-to-air missiles (Magic
        missiles on Indian Navy Sea Harriers) or
        a combination of air-to-surface missiles,
        bombs, rockets or two 30 mm guns

stands at level three overall, with both sides operating with considerable risk in time of war. However, taking into account the West's desperate need to reinforce Europe during a period of rising international tension or outright hostilities, requiring sea lines of communications (SLOCs) across the Atlantic to be defended by all-too-few escort vessels such as frigates and destroyers, the odds favour any attacker possessing a massive fleet of submarines and missile cruisers.

In the past, NATO has tended to envisage situations which are essentially defensive or reactive to threat – defending the SLOCs and, at best, bottling up enemy surface and sub-surface forces north of the so-called Greenland–Iceland–UK (GIUK) gap, or chokepoint, before reinforcement shipping could be provided. Today there is a growing belief on both sides of the divide that the best way to counter the threat at sea is to take the battle to an enemy. As a US Navy admiral has stated: 'In war at sea, the most rapid, efficient and sure way to establish the control of essential sea areas, is to destroy the opposing forces capable of challenging your control of those seas.' The CBG has become the deciding factor in the maritime balance of forces, because it enables the swift projection of friendly forces into an area that can be comfortably controlled. For NATO this would perhaps mean the reinforcement of Norway by US, British and Dutch Marines, to protect its vital northern flank.

The Soviet Navy, for example, has a large and still-growing capability to prevent the West's reinforcement attempts, while it is now also able to use its aircraft carriers in conjunction with powerful nuclear-powered battle cruisers and other escorts for intervention in overseas problem areas during peacetime (a role familiar to US Navy carriers), and other more conventional tasks in war. Although the lack of large aircraft carriers within the Soviet Navy has been cited in the past by some people in the West as evidence that resources have been expended unnecessarily on vessels of this type, Soviet acceptance of the necessity for 'organic air' does predate 1963, which is when the first *Moskva* class anti-submarine helicopter cruiser/carrier was laid down. *Moskva* itself was commissioned in 1967 and was joined by *Leningrad* late the following year, each vessel carrying 18 Kamov Ka-25 Hormone-A helicopters with anti-submarine warfare (ASW) as their main function. The newer *Kiev* class aircraft carriers extend the anti-submarine role but are classified as tactical aircraft-carrying cruisers. The development of a VTOL warplane, the Yak-38 Forger, gave Soviet ship designers cause to think in terms of carriers for fixed-wing aircraft, resulting in the *Kiev* class. Although fewer than half its complement of an estimated 32 aircraft are normally Forgers, a *Kiev* class carrier can also carry surface-to-air missiles (SAMs), anti-shipping cruise missiles, anti-submarine weapons, guns and torpedoes. Now huge Soviet nuclear-powered aircraft carriers are under construction, and these will be discussed later.

Yakovlev Yak-38 (USSR)

The only Soviet V/STOL fixed-wing aircraft in operational service, Forger (NATO name) is a carrier-borne combat aircraft found in Forger-A single-seat form and Forger-B two-seat training form on board *Kiev* class vessels, 12 and 1 in each vessel respectively. It is assigned the task of attack and reconnaissance, for which it has ranging radar in the nose and four underwing pylons to which to attach stores. Like the Sea Harrier, Forger can perform both vertical and short take-offs on-board ship, although the Forger's use of the reportedly less efficient vertical-lift method of using two separate liftjets in the forward fuselage and a main horizontal engine with one pair of vectoring nozzles makes it less versatile. For short take-offs, a system automatically controls the lift engines and rotates the vectoring side nozzles during the short run, providing a very smooth take-off with the best payload. The automatic control system comes into use again for the vertical landing. Although clearly of far greater combat potential than once considered likely in the West, it is thought unlikely that Forger will have a place on the new Soviet nuclear-powered carriers now under construction.

Data: Forger-A, estimated
Number in crew: One
Engines: One 17,990 lb (8160 kg) thrust Lyulka AL-21 turbojet exhausting through two side vectoring nozzles. Two 7880 lb (3575 kg) thrust lift engines mounted vertically in tandem
Length: 50 ft 10 in. (15.50 m)
Wing span: 24 ft 0 in. (7.32 m)
Gross weight: 25,795 lb (11,700 kg)
Maximum speed: Nearly Mach 1
Combat radius: up to 230 miles (370 km)
Weapons: Possibly up to 7936 lb (3600 kg) of weapons, including AS-7 Kerry air-to-surface missiles, Aphid air-to-air missiles, other missiles, bombs, rockets or gun pods

▶

An artist's impression of Kremlin *under construction at the Nikolayev 444 shipyard.*

What is a CBG?

In US Navy terms, a CBG consists of one or two aircraft carriers, each carrying typically 86 aircraft, protected by other surface ships with a particular role. The *Ticonderoga* class of cruiser, equipped with the Aegis system of air defence based on SPY 1A phased array radar, provides a near-impenetrable protective umbrella against air attack on the CBG. *Spruance* class destroyers and *Oliver Hazard Perry* class frigates provide anti-submarine protection with their SH-60B Seahawk helicopters throwing a screen around the CBG, dipping (or 'dunking') sonar on the end of a cable below the surface of the sea in order to detect noise generated by any lurking enemy submarine. If located, a small task force of ships, carrier-borne S-3A Viking ASW aircraft and rotary aircraft is despatched to attack the submarine with depth charges and torpedoes. Defence against surface attack is maintained by the carriers' strike aircraft: the F/A-18 Hornet is now replacing the A-7E Corsair II in the light attack role, but the A-6E Intruder continues in service for medium attack duties. These are backed up by the Harpoon anti-ship missiles of the CBG escorts, although the Tomahawk cruise surface-to-surface missile (SSM) is being retrofitted to *Spruance* class destroyers.

Thus, between six and ten ships form a CBG, including replenishment tankers and stores ships for refuelling and rearming the combatants. Although this sounds a potent force, very much in the striking power league of the battleship task forces of the Second World War, it is not a tightly-bunched naval group steaming into battle. With the aid of modern radar and sonar sensors, a CBG can disperse itself over perhaps 60 nautical miles, each element probing the air and sea for any sign of hostile presence with, far ahead, the carriers' E-2C Hawkeye aircraft providing long-range warning of attack.

Despite the layering of defences and perhaps the grouping of three or even four carriers, losses are likely to occur in a conventional war, but naval analysts estimate that between 30% and 40% of a carrier force could be lost in a nuclear weapons exchange at sea. Even with 15 US Navy CBGs, there are too few forces available to meet too many commitments. Routinely, a CBG of six surface combatants is detached from the US 7th Fleet in the western Pacific for duty in the Indian Ocean; one or two carriers are always present in the Mediterranean and six in the Pacific. In a period of tension or actual hostilities, US Navy and NATO commanders could find it difficult to allocate forces for their particular area, with an order of priority having to be decided. All this leaves precious little for reserves, which suggests that, at the least, extreme caution should be exercised by politicians in deciding whether to risk a US carrier 'national asset'.

At present, an enemy's surface groups would become vulnerable to attack by NATO carrier-borne strike aircraft once they had moved out of range of their own land-based air support. Conscious of this the Soviet Navy, for example, has developed a two-pronged method of evening out the odds. The first prong is the establishment of forward operating bases for Soviet Naval Aviation aircraft: Tupolev Bear reconnaissance and anti-submarine aircraft now regularly fly out of bases

◄

The second Soviet Navy Kiev-class aircraft carrier Minsk, *with a Hormone helicopter on the forward deck and Forgers aft.*

▷ ▲

The Falklands conflict of 1982 proved the value of sea control using carrier air power, typified by this remarkable photograph of a Sea Harrier and Sea King operating from HMS Hermes.

▷

The F/A-18A Hornet offers up a new shape on the US carrier deck.

Tupolev Tu-142 Bear-D

1 Fixed flight-refuelling probe
2 Observer's compartment nose glazing
3 'Short Horn' J-band frequency-agile navigation and bombing radar
4 Avionics equipment bay (port and starboard)
5 'Odd Rods' IFF aerials
6 Flight deck enclosure
7 Nose undercarriage pivot mounting
8 Retractable landing/taxiing lamps (port and starboard)
9 Nosewheel steering jacks
10 Aft-retracting twin nosewheels
11 Nosewheel doors
12 Pitot tubes
13 Flight-deck roof escape hatch
14 Forward pressurized crew compartment

15 Observation dome
16 Wing-root attachment joint
17 Wing centre-section carry-through
18 Inboard wing panel
19 Starboard engine nacelles
20 AV-60N eight-bladed contra-rotating propellers
21 Propeller spinners
22 Wing fences
23 Outboard wing panel
24 Wing-tip lighting
25 Starboard aileron
26 Aileron tab

27 Outboard Fowler-type flap (lowered)
28 Flap guide rails
29 Nacelle tail fairing
30 Extended tail fairing (some Bear-F aircraft)
31 Mainwheel doors
32 Inboard Fowler-type flap (lowered)
33 Satellite communications antennae
34 HF communications aerial
35 Circular-section unpressurized fuselage

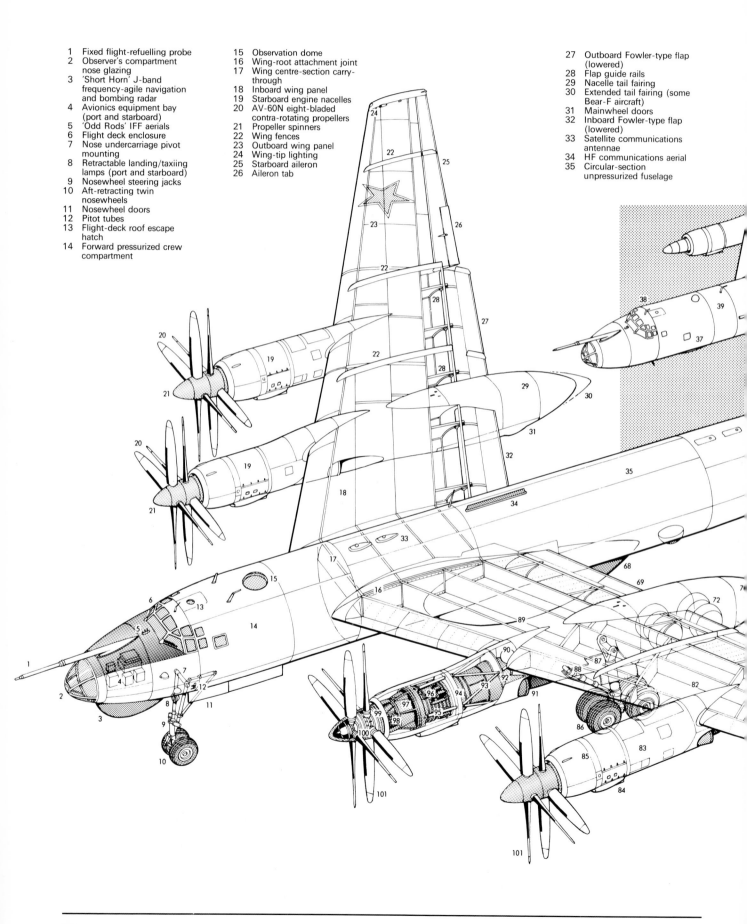

36 Retractable dorsal barbette (twin 23 mm NR-23 cannon) usually faired over
37 Additional observation port (Bear-E and -F)
38 Raised flight-deck roof (Bear-F)
39 Extended crew compartment fuselage plug (Bear-F)
40 Search radome of Bear-F (smaller and set further forward than 'Big Bulge' of Bear-D)
41 Two-section stores stowage bay (Bear-F)
42 Stores bay camera pallet (Bear-E)
43 Fin root fillet
44 Starboard tailplane
45 Tailfin
46 VOR ground control communications antennae

47 HF aerial cable
48 Fin-tip VHF aerial fairing
49 Magnetic anomaly detector (MAD) boom (some examples of Bear-F)
50 Rudder
51 Rudder tab

52 Sensor equipment tail fairing (replacing tail barbette on some examples of Bear-D)
53 'Box Tail' or 'Bee Hind' I-band tail warning radar (working in conjunction with 55)
54 Tail gunner's compartment
55 Paired 23 mm NR-23 cannon
56 Elevator trim tab
57 Port elevator
58 Tailplane tip sensor fairing
59 Port tailplane
60 Observation blister (port and starboard)
61 Ventral twin 23 mm NR-23 cannon barbette (deleted from some aircraft)
62 Lateral dielectric blister
63 Camera port

64 Fuselage profile of Tu-95 Bear-C strike aircraft
65 Semi-recessed missile housing
66 AS-3 'Kangaroo' air-to-surface missile
67 'Crown Drum' nose radome

68 Ventral 'Big Bulge' I-band search radar
69 Port inboard Fowler-type flap
70 Nacelle tail fairing
71 Extended tail fairing (some Bear-F aircraft)
72 Main undercarriage stowed position
73 Outboard Fowler-type flap
74 Aileron tab
75 Port aileron
76 Wing-tip fairing
77 Port wing-tip lighting
78 Leading-edge thermal de-icing
79 Wing fences
80 Outer wing panel (three-spar construction)
81 Outboard wing panel joint rib
82 Wing integral fuel tanks (total fuel capacity 16,540 Imp. gal./72,980 litres)
83 Port outboard engine nacelle
84 Ventral oil cooler
85 Engine cowling panels
86 Four-wheel aft-retracting main undercarriage bogie
87 Main undercarriage leg strut
88 Hydraulic retraction jack
89 Inboard wing panel (four-spar construction)
90 Engine fire extinguisher bottle
91 Exhaust duct
92 Bifurcated jet pipe
93 Engine bearer struts
94 Main engine mounting ring frame/firewall
95 Kuznetsov NK-12MV turboprop now uprated to (14,795 e.h.p.)
96 Engine accessory equipment
97 Engine air intake
98 Propeller reduction gearbox
99 Engine cowling annular air intake
100 Propeller hub pitch-change mechanism
101 Port contra-rotating propellers

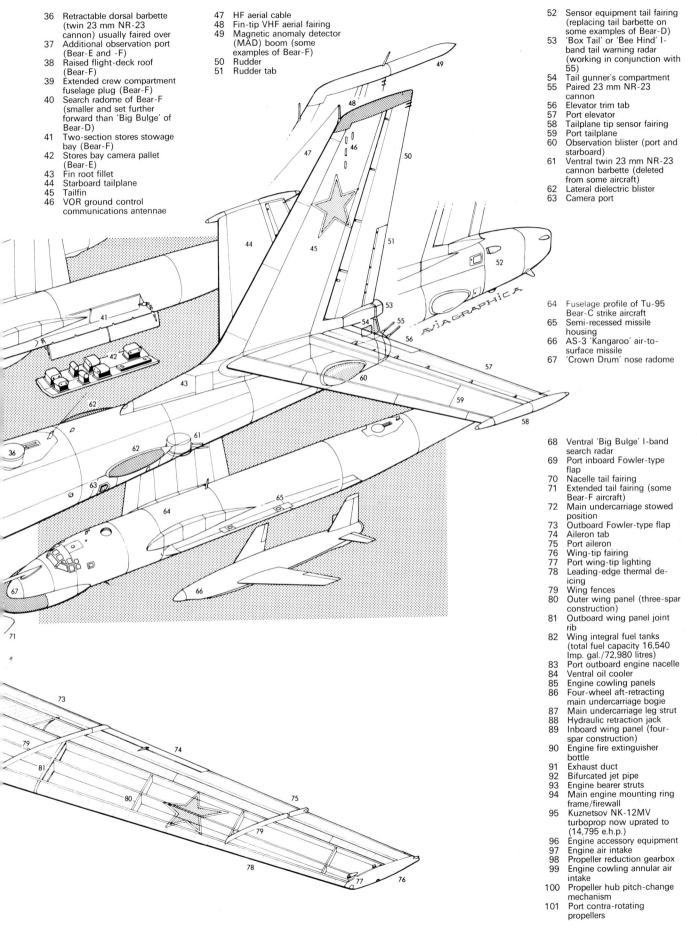

in Cuba, Angola and the former US base at Cam Ranh Bay in Vietnam, providing a protective umbrella (if necessary) for any naval operations planned for those regions. Allied to this are moves to extend the range of land-based aircraft, particularly for those operating out of airfields in the Kola peninsula around the Arctic Circle in the north-west of the USSR. In time of war, the Soviet High Command must be expected to launch an offensive against northern Norway, to capture the Norwegians' four or five vital military airfields for use by Soviet Naval Aviation aircraft so as to extend the range of operations further out into the Norwegian Sea and the North Atlantic. The Soviets are also believed to be developing a new long-range fighter, capable of a lengthy loiter time,

▲▲
CBG Third Fleet Task Force 177's Group Delta under way following a mobile sea range exercise. The ships are (from left) USS Berkeley guided-missile destroyer, USS Truxtun nuclear-powered guided-missile cruiser, USS Halsey guided-missile cruiser, USS Sterett guided-missile cruiser, USS Kitty Hawk aircraft carrier, USS Bainbridge nuclear-powered guided-missile cruiser, USS William H. Standley guided-missile cruiser and USS Henry B. Wilson guided-missile destroyer.

▲
An E-2C Hawkeye returns from an airborne early warning patrol.

▲
Tupolev Tu-142 Bear-D.

▶
Soviet naval reconnaissance aircraft operating areas in 1985, according to US sources. A black spot marks newly constructed airfields.

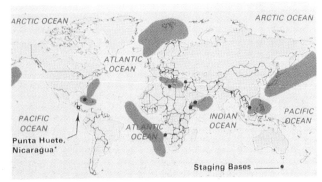

FLEXAR

FLEXAR is a shipboard weapon control system, proposed to provide the US Navy with a complement to its existing radars. It is designed to search high and low elevation sectors to detect, track and bring defensive fire upon pop-up targets close to the surface, or others diving from high angles, its fast reaction time being essential to allow intercept of high-speed targets suddenly appearing at short ranges. FLEXAR's antenna can also be stopped to engage multiple targets in dense threat sectors over 120° in azimuth, providing illumination for semi-active missiles while maintaining target tracking and continuing the search for new targets (Hughes Aircraft).

solely to counter carrier-borne aircraft and also to disrupt NATO's 'air bridge' across the Atlantic. This aircraft, regularly refuelled by air-to-air tankers, would be able to intercept the requisitioned civilian wide-bodied jets bringing personnel and light stores (such as ammunition) from the USA to Europe during the early stages of hostilities. To give some idea of the scale of the reinforcement operation envisaged, one reinforcement/resupply jet would fly over a man standing on the bridge of a ship in mid-Atlantic every three minutes.

As already mentioned, the Soviet Navy decided some years ago that it too would move into the realm of 'organic air' at sea, part of the process that has transformed the Soviet Navy from merely a coastal defence force at the end of the Second World War to a world power. There is little doubt that the Cuban missile crisis of 1962 hastened the

construction programme, which has consisted of three well-defined phases. First there were the two helicopter cruisers *Leningrad* and *Moskva*, intended purely as anti-submarine warfare (ASW) ships. Then came the *Kiev* class carriers, with their complement of Yak-38 Forger jump-jets for localized air defence and light strike duties against surface targets and a number of Kamov Helix helicopters for ASW, communications and search-and-rescue work. Finally, there is the nuclear-powered strike carrier which is now under construction at the Nikolayev 444 shipyard on the Black Sea. This 1100 ft (335 m) long, 75,000 ton ship, the first of perhaps eight carriers of the class, will be able to accommodate 60–70 aircraft and will begin sea trials in 1988, with operational service planned for early 1994. The first vessel, *Kremlin*, will probably go to sea with a navalized version of the MiG-23 Flogger on

*Kamov Ka-27 Helix-A on board the third Kiev-class carrier,
Novorossiysk.*

board, but eventually the new Sukhoi Su-27 Flanker will
probably become the main carrier aircraft. Already trials
involving aircraft on a concrete mock carrier deck have
been observed at an airfield near the Black Sea.

The Flanker, comparable with the US F-15 Eagle, is
armed with eight AA-10 radar homing missiles and has
recently entered service in land-based form. Its two
engines, each developing 30,000 lb (13,600 kg) thrust with
afterburners, provide a maximum speed of Mach 2.35 in
level flight. Western intelligence sources believe its
combat radius is in the region of 620 nautical miles.
Reported problems with engine production have delayed

deliveries of aircraft being built at Komsomolsk in the
Soviet Far East, although the same type of powerplant is
thought to be fitted to the MiG-31 Foxhound which
recently entered service with Soviet Air Force units in
East Germany.

The *Kremlin*, like the vessels in the *Kiev* class, bristles
with armaments and sensors. It is believed that it will be
equipped with six silos for vertically launched surface-to-
air missiles positioned just forward of the island
superstructure. Above the bridge, phased array radars
similar to the SPY-1A systems of the US Navy's Aegis
system will be installed. Three deck lifts are allocated for
the movement of aircraft and three steam catapults are
provided for launching the strike planes. Once operational, the *Kremlin* will form the core of a formidable
CBG, particularly if grouped with a member of the *Kirov*

Ship-borne missiles for area defence

Ship-borne missiles for defence against aircraft date from the 1950s. The US Navy, for example, first deployed the Convair Terrier in 1956, and since then SAMs have been essential to warships of frigate size and larger but can be fitted to smaller vessels if required. In recent years a new threat against ships has appeared in the form of highly accurate, sea-skimming anti-ship missiles.

Weapon systems designed to deal with both threats are deployed by navies around the world. For the Royal Navy, the latest area defence missile is Seawolf, which has all the necessary attributes for success: it is automatically on guard, quick to react to a threat and has quick acceleration to the intercept point. In operation, a lightweight radar tracks the target and the Seawolf missile, which is guided to the intercept point by command to line-of-sight. Advanced differential tracking techniques are used to ensure high 'kill' probabilities against both missile and aircraft targets. Seawolf is controlled throughout its flight by a microwave command link. To maximize reliability and reduce cost per round, guidance calculations are carried out on board ship so that a minimum of electronic equipment is carried in the missile. When required, a salvo of two missiles can be fired under the control of a single radar.

The photograph shows Seawolf on board a Royal Navy Type 22 frigate (British Aerospace Dynamics).

Kara-*class missile cruiser, intended primarily for ASW.*

class nuclear-powered battle cruisers.

The Soviet Navy has recently begun practising CBG tactics with their carriers, a major development in the maritime balance of power. In April 1985 a CBG group, centred on the carrier *Novorossiysk*, left the Soviet Navy's Far East base at Vladivostok for a high-speed exercise in the Pacific. Eight ships were in the task force; three *Kara* class cruisers, a *Kresta II* cruiser and two *Krivak* frigates – two replenishment ships completed the group. During the exercise, conducted at the surprisingly high average speed of 14–18 knots throughout, the *Krivak* frigates sailed between 20 and 30 nautical miles ahead of *Novorossiysk* with the three *Kara* cruisers keeping station 4 and 11 nautical miles apart – a classic CBG deployment.

After circling Japan, the Soviet naval force, shadowed by Japanese Maritime Self-Defence Force (JMSDF) ships and aircraft, returned to Vladivostok, having demon-

strated that Soviet CBGs are now a force to be reckoned with.

Two months later there was a second demonstration. The sister ship *Kiev*, having completed a refit in the Black Sea, sailed out of the Mediterranean and began operating as part of a CBG as it headed north to rejoin the Soviet Northern Fleet, based in Murmansk. In company with it were two *Kresta* cruisers, three *Sovremenny* destroyers, one *Modified Kashin* destroyer and a *Krivak* frigate, plus a tanker.

During the passage, *Kiev* was observed by shadowing NATO forces to be launching its Yak-38 Forger jump-jets in a new manner. Previously, these had taken off vertically with the resulting penalties in weaponry payload. Now the aircraft were being launched using rolling take-offs, very similar to the method used by the Royal Navy's Sea Harriers and with less fuel consumed. The Forgers conducted bombing and gunnery exercises against a towed target during the passage and anti-submarine techniques were also rehearsed.

▶

Kirov-class nuclear-powered battle cruiser, equipped with large numbers of surface-to-surface anti-ship and surface-to-air missiles, guns, torpedoes, anti-submarine weapons and three Hormone or Helix helicopters.

The scale of interest in this new Soviet tactic was signified by the sailing of the US Navy's nuclear-powered cruiser, the 10,000 ton *Mississippi*, from Norfolk, Virginia. After a 30 knot dash across the North Atlantic in a force-eight gale, she joined the NATO shadowing warships off the coast of Norway and watched as Soviet land-based aircraft exercised with the *Kiev* CBG as it headed for Murmansk. But the power of the embryonic Soviet CBGs is limited by a lack of practical experience in the operation of aircraft carriers, even though the operation of the West's carriers has been carefully monitored over a period of 25 years.

In a continuing effort to improve its capabilities, the Soviet Navy conducted a massive exercise in the Atlantic Ocean, North Sea and Norwegian Sea in July 1985, involving aircraft, more than 50 surface warships and about 32 submarines. Intended to simulate the reinforcement by NATO of its northern flank and the Soviet response forces, the vessels were drawn from the Baltic, Black Sea and Northern Fleets. As expected, the main CBG force was headed by *Kiev*, accompanied by *Kirov* and escorted by two *Sovremenny* class, two *Udaloy* class and a *Kashin* class destroyers, and two *Kresta* cruisers. The exercise was followed closely by NATO powers, the RAF intercepting many more Soviet aircraft in the UK air defence region than usual. Indeed, the exercise was the largest involving co-ordinated Soviet maritime forces in modern history, and included around 300 air sorties.

Although it is true that several countries are able to form CBGs, only those of the USA and USSR are sufficiently powerful and have the political backing to be used to influence events across the world in peacetime, their awesome weaponry and air power being capable of actually preventing hostility when deployed off the coast of a trouble spot. Whilst carrier aircraft have been used operationally many times since the Second World War, including during the Korean, Vietnam and Falklands conflicts, modern CBGs have yet to face each other in direct hostility as in the famed carrier battles of the Pacific War with Japan. The 'flat-tops' within a surface force will remain the priority naval targets in war, and their survival in a conventional conflict will depend at least as much on the escort ships, and on the electronics-carrying aircraft and defence weapons on board, as on the ability of the carrier fighters and strike aircraft to knock out potential aggressors. A comment by a US Navy carrier commander may equally have been spoken by a Soviet commander, when he said '. . . but a nuclear-powered strike carrier with all that weaponry on board will be damned hard to take out'.

▶

Sovremenny-class destroyer, its single Helix almost certainly being used for target acquisition purposes for the eight SS-N-22 anti-ship missiles. A telescopic hangar is provided for the helicopter.

The first Slava-class cruiser was commissioned into the Soviet Navy in 1982, its 16 SS-N-12 Sandbox anti-ship missiles being aided by a Helix-B target acquisition and mid-course guidance helicopter. Helix-B may also be capable of ASW with torpedoes and depth bombs or, for this role, a Helix-A might be substituted.

Coastal Defence by missile

Some nations with long coastlines employ shore-based versions of anti-ship missiles in coastal defence roles. Such missiles are usually truck mounted for mobility and are launched from ramps or salvo containers. Supporting vehicles carry radar and other equipment.

The photograph shows a night firing of the French MM40 version of Exocet from a four-missile launcher/container. The drawing depicts a Norwegian Penguin coastal defence system, comprising the launcher vehicle with missiles and the surveillance radar vehicle which provides target detection and designation. MM40 has a range of 43.5 miles (70 km) and Penguin 18.5 miles (30 km). Credits: MM40 (Aérospatiale) and Penguin (Kongsberg Vaapenfabrikk).

▶

The bridge of USS Nimitz, with a harbour pilot in addition to the normal crew during docking at a naval base.

▼

Combat information centre on-board USS Nimitz.

▷▶

An aircraft carrier has to be a totally self-contained community, with all essential services. Here a hospital corpsman readies equipment for use in a medical laboratory on board ship.

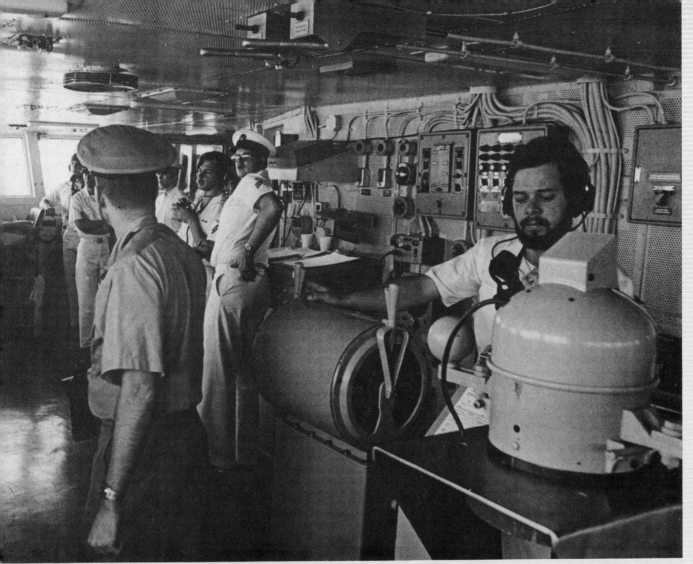

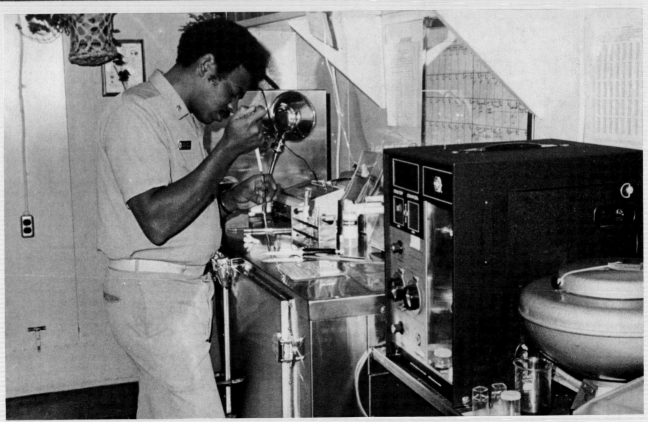

Chapter 3
Air power
and the
sub-sea threat

With the development of nuclear weapons the submarine, long the feared enemy of ships' captains, took on a new role. From deep below the ocean surface, a submarine could now launch a powerful nuclear missile with an accuracy certainly good enough to wipe out any target specified, even thousands of miles away. At the same time, the ability of 'conventional' torpedo-equipped submarines to detect and destroy surface ships greatly increased, in return making the submarine a prime target in war. But, just like a needle in a haystack, a submerged submarine is very difficult to find and, not unnaturally, submarine designers do all they can to camouflage the vessels to make detection even more difficult.

Given their tremendous latent destructive power, it is natural in these days of sophisticated measure and countermeasure that the equipment for detecting and destroying submarines has developed dramatically over the past 50 years. Today anti-submarine warfare (ASW) forms a major element in any naval force, although only those nations facing the severest threat employ the more sophisticated countermeasures. During the massive Soviet naval exercise of July 1985, known to NATO as Summerex 85, one-third of all air sorties were flown by Bears and Mays dedicated to ASW.

There are three main categories of anti-submarine

Preceding pages

Soviet nuclear powered Victor III-class fleet submarine, fitted for torpedoes and SS-N-15 anti-submarine missiles.

Soviet nuclear-powered Delta III strategic submarine, with 16 launch tubes for SS-N-18 ballistic missiles in addition to torpedo tubes.

aircraft to consider: those for long- and medium-range maritime patrol (the albatrosses of the aviation world), the short-range carrier-borne aircraft, and the ship-borne helicopter. Nations with large coastlines to defend against sophisticated sub-sea attack, ranging from Australia and Japan in the east to the countries of Europe, all have at least one of these types in their inventories, and, not surprisingly, the USA and the USSR are equipped to defend themselves and their allies against the most extreme threats posed by the opposition's nuclear submarines.

Long-range aircraft

Long-range maritime aircraft are typified by the British Aerospace Nimrod M.R. 2, in service solely with the Royal Air Force, the Lockheed P-3 Orion, in service with the US Navy and nine other nations around the world, and the Soviet Tupolev Bear which patrols not only from Warsaw Pact bases but from countries such as Cuba and Angola, plus the Ilyushin Il-38 May. These types have two key features in common: they have very long patrol endurances – at least ten hours, since finding and tracking submarines is a long and arduous task – and they possess large airliner-like cabins to house the crew (about six observers is normal) and sophisticated electronics needed to perform their task; indeed all these types are based on civil airliners of a previous generation, the Comet, Electra, Tu-114 Cleat and Il-18 respectively.

Nations requiring smaller twin-engined ASW/patrol aircraft, or perhaps wishing to complement a fleet of four-engined types, can select potent aircraft such as the French Dassault–Breguet Atlantic/Atlantique, the former as currently used by France, West Germany, Italy, the Netherlands and Pakistan. Such aircraft are about as capable as the larger types but tend to patrol closer to

Dassault-Breguet Atlantic (France)
The Atlantic is a maritime patrol aircraft that was in production for a decade from 1964. Intended originally as a Lockheed P-2 Neptune replacement with NATO, it was ordered only by France, Germany, Italy and the Netherlands, although Pakistan subsequently took over three of France's delivered 40. Of Germany's Atlantics, five are currently operated as special electronic intelligence aircraft, the remaining 14 operating the traditional role. A new version of the aircraft has entered production as the Atlantique 2, but deliveries will not begin until 1989.

Number in crew: 12
Engines: Two 6106 e.h.p. SNECMA/Rolls-Royce Tyne
    RTy 20 Mk 21 turboprops
Length: 104 ft 2 in. (31.75 m)
Wing span: 119 ft 1 in. (36.30 m)
Gross weight: 95,900 lb (43,500 kg)
Maximum speed: 409 mph (658 km/h)
Range: 5590 miles (9000 km)
Weapons: Weapons bay for: bombs, depth charges
    and torpedoes
    Underwing attachment points for: four air-
    to-surface missiles or rockets

home bases. Several countries use second-hand refurbished aircraft as a cost-effective way of monitoring sub-sea activity (Pakistan has received three Atlantics from France, for example). More unusual are flying-boats, a type of aircraft that lost favour post-war as land-based aircraft obtained greater ranges. Japan's JMSDF received Shin Meiwa PS-1s for ASW, but by far the world's major operator of anti-submarine/maritime patrol flying-boats is the Soviet Navy with its fleet of Beriev M-12s (Be-12s). M-12s are standard equipment at Northern and Black Sea Fleet coastal bases but normally venture no further than 200 nautical miles from shore.

The USA is alone in having a modern fixed-wing carrier-based ASW aircraft in the Lockheed S-3 Viking, for while the land-based P-3 has long legs, the ASW role of protecting CBGs far from shore is best left to a specialist aircraft permanently on station. There are other carrier-based ASW aircraft, including the French Alizé serving both France and India, but these are of older design.

The main anti-submarine system is the sonobuoy, and all the above types of aircraft are primarily equipped with sonobuoy receivers and analysers. There are two types of buoy; active and passive. Both float in the sea beneath a flotation chamber (which encloses the transmitting antenna) and listen for under-sea noises. The passive buoy simply floats and listens, and then re-transmits what it hears to the patrolling aircraft when triggered to do so. The active buoys are similar to radar in that they send out sonic beams and listen to the reflections from under-sea objects (some units can be both active and passive in order to increase operational flexibility).

Buoys have controls to allow them to 'float' at a variety of depths and to operate for a specified length of time at each one. When triggered to respond to a co-operating

aircraft, modern buoys can transmit on a range of 99 frequency channels, to help avoid being overheard by an enemy's detectors. At the end of the specified operational time the buoy remains dormant for some hours and then a chemical reaction is activated which results in the buoy sinking to the bottom of the sea, thus preventing an enemy capturing it and analysing its electronics.

Of course, submarines are not alone in making noise under water, but they are very carefully designed to be as quiet as possible – indeed in-service submarines are regularly tested on special ranges to make sure they have not become any noisier, perhaps because some machinery has become slightly loose or requires lubrication. The ASW systems designer, therefore, has the increasing

Beriev M-12 Tchaika (USSR)
Known to NATO as Mail, the M-12 (or Be-12) is an unusual looking maritime patrol amphibian of which about 90 are operated by the Soviet Naval Air Force from bases of the Northern and Black Sea Fleets. Their main roles are maritime surveillance and anti-submarine, for which search radar is carried in a nose 'thimble' radome and MAD in a tail 'sting'.

Engines: Two 4190 e.h.p. Ivchenko AI-20D turboprops
Length: 99 ft 0 in. (30.17 m)
Wing span: 97 ft 6 in. (29.71 m)
Gross weight: 64,925 lb (29,450 kg)
Maximum speed: 378 mph (608 km/h)
Range: 2485 miles (4000 km)
Weapons: Torpedoes, depth charges, sonobuoys and
    other stores in the bay and under the
    wings

Ilyushin Il-38 (USSR)
The Il-38, known to NATO by the code name May, is the Soviet Naval Air Force's standard shore-based anti-submarine and maritime patrol aircraft, equivalent to the US Navy's Orion but based upon the Il-18 airliner airframe. Fifty to sixty are in Soviet use and three with the Indian Navy, each carrying search radar in an underfuselage radome, MAD in a tail 'sting' and much other equipment. May is mainly seen patrolling the Atlantic and Mediterranean but now often operates also over the Indian Ocean and the Red Sea.

Number in crew: 12
Engines: Four 4200 s.h.p. Ivchenko AI-20M turboprops
Length: 129 ft 10 in. (39.60 m)
Wing span: 122 ft 9 in. (37.42 m)
Gross weight: 140,000 lb (63,500 kg)
Maximum speed: 400 mph (645 km/h)
Range: 4473 miles (7200 km)
Weapons: Full range of anti-submarine weapons

▲
The RAF's long-range maritime patrol aircraft is the Nimrod MR.2, carrying advanced search radar, early warning support measures (EWSM) in wingtip pods (when fitted), electronic support measures (ESM) in a pod on the fin, MAD in a tail 'sting' and other equipment.
◀
The GEC AQS-901 acoustics processing and display system on-board a Nimrod MR.2, which is compatible with all NATO passive and active sonobuoys.
▷ ▲
Dassault-Breguet Atlantique with bay doors open to reveal a typical internal weapon load of three Mk 46 homing torpedoes and an AM 39 Exocet anti-ship missile.
▷ ▼
The standard Soviet ASW and maritime patrol aircraft is the Ilyushin Il-38 May

Soviet Beriev M-12 Tchaika photographed by the Royal Danish Air Force while on patrol.

difficulty of trying to analyse less and less submarine-created noise amid all the under-sea background disturbances.

Passive buoys are most commonly used; they are cheaper, dispensable and, most importantly, because they are passive they cannot be detected by the enemy's electronic surveillance equipment. An array of buoys will be laid in the area between the likely position of enemy submarines and the land being defended, or just at strategic places in the ocean. The buoys are launched from an aircraft through an exit tube in the rear fuselage – an Orion can carry up to 87 buoys and can monitor 31 at any one time, for example, while the Atlantique 2 carries more than 100.

Having deployed the required number of sonobuoys, the aircraft flies close to the array from time to time (it may have several groups of buoys to monitor), receiving from each buoy what that buoy is hearing in the form of a noise spectrum and a direction indication. A typical buoy will be able to transmit its data to an aircraft 10 miles (16 km) or more away. The ASW computer on the aircraft will then analyse the noise inputs and determine the direction of any possible threat in relation to each buoy. Simply by correlating all the directional information, and by applying Doppler techniques or monitoring the noise over a period of time, the exact location, track and speed of any submarine in the area can be determined quite accurately. The computer will carry a library of specific submarine noise signatures, built up over years from many observations, so that it is often possible for the aircraft's observers to name each submarine they detect, simply by listening to the noise it emits.

Active buoys are used more sparingly, for although they provide more accurate information they are detectable by the target submarine, which can then launch a counter-attack. In both cases the systems designer's job is a very difficult one, and the latest computer technology and signal processing techniques have to be employed. Data are presented to the crew on television screens or as 'hard copy' (paper print-outs) for future analysis. Once a submarine has been identified, its most recent signature is stored in the computer and is made available for the rest of the nation's ASW forces.

For long-term monitoring of strategic areas, such as the Greenland–Iceland–UK gaps, permanent arrays of hydrophones are laid on the sea bed which are regularly monitored by ships (in this case those of NATO). Surface vessels have very extensive ASW capabilities; aircraft are used more in the tactical role, monitoring relatively large areas of sea at frequent intervals rather than the smaller areas around a ship all the time.

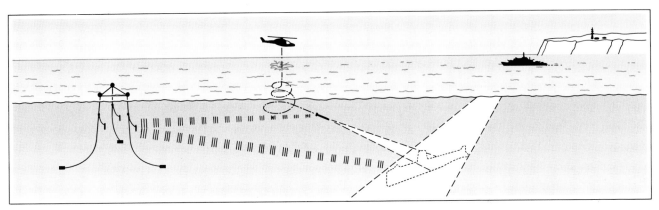

A portable acoustic system; hydrophones are suspended from surface floats and held in position by buoys; data are relayed to a control station (© Thorn EMI).

Helicopters

The protection of a naval surface force is a vital task, as was amply demonstrated in the 1982 Falklands war between Britain and Argentina, and although ships have very sophisticated ASW devices on board, ASW helicopters flying as sentries or pickets ahead of the fleet can greatly enhance this detection capability. Again, the major maritime nations all have extensive anti-submarine helicopter fleets, both shore based and deployed on-board aircraft carriers, and operating in ones or twos from many cruisers, destroyers and frigates.

The main equipment used by these helicopters is dipping sonar. A large passive sonar, with very sensitive detecting arrays, is lowered into the sea from the hovering helicopter, and the sounds detected are passed up the connecting cable for immediate analysis. Given that a

▼
The Soviet Mil Mi-14 Haze-A ASW helicopter employs a MAD towed 'bird' from behind the main cabin.

helicopter can move far quicker than a submarine, repeated 'dunkings' (as these sonar dippings are called) over a wide area will soon build up a comprehensive picture of under-sea activity around the coast or fleet.

Magnetic anomaly detector

Sonar is the main sensor carried in aircraft for underwater detection, but aircraft also carry a comprehensive suite of electronic survey sensors to maximize their effectiveness. Perhaps the next most-important after sonar is the magnetic anomaly detector (MAD), which is most noticeable in aircraft such as the Nimrod, Atlantique, Il-38 and Orion where the system is contained in a long boom at the tail of the aircraft (referred to as a tail 'sting'). MAD also equips helicopters, particularly small ones which cannot carry a full ASW suite.

The MAD equipment, as its name implies, detects any anomalies in the earth's magnetic field. Apart from the naturally occurring magnetic field, which varies in strength and direction all over the earth's surface, the aircraft itself creates a sizeable electro-magnetic disturbance, largely caused by the electronics in the cabin. Because MAD attempts to detect anomalies created by submarines some distance below the surface of the sea,

◄
The Orion, in updated versions, has been the standard shore-based maritime patrol aircraft of the US Navy since 1962.

▷
Royal Navy Westland Sea King HAS.5 winching its Plessey Type 195 dipping sonar from the sea.

▼
The unique carrier-based S-3A Viking.

the equipment has to be very sensitive and situated as far from the aircraft's other electronics as possible – hence the use of the tail boom in fixed-wing aircraft. In helicopters the MAD detector is usually towed behind on a long cable, to minimize the helicopter's effects, but recent advances in signal processing now make it possible to have an installation fixed to the cabin.

In moving through the water the submarine creates its own magnetic field, which the sensitive detectors can resolve from the other magnetic effects. However, as is the case with noise, submarines are designed to have as little magnetic effect as possible and are regularly tested to ensure that no unusual signatures are created by faulty machinery (for example, even a dent in the hull caused by

Sikorsky SH-60B Seahawk

1 Graphite/epoxy composite tail rotor blades
2 Lightweight cross beam rotor hub
3 Blade pitch-change spider
4 Anti-collision light
5 Tail rotor final drive bevel gearbox
6 Rotor hub canted 20°
7 Horizontal tailplane folded position
8 Pull-out maintenance steps
9 Port tailplane
10 Tail rotor drive shaft
11 Fin pylon construction
12 Tailplane hydraulic jack
13 Cambered trailing-edge section
14 Tail navigation light
15 Tailplane hinge joint (manual folding)
16 Handgrips
17 Static dischargers
18 Starboard tailplane construction
19 Towed magnetic anomaly detector (MAD)
20 Tail bumper
21 Shock absorber strut
22 Bevel drive gearbox
23 Tail pylon latch joint
24 Tail pylon hinge frame (manual folding)
25 Transmission shaft disconnect
26 Tail rotor transmission shaft
27 Shaft bearings
28 Tail pylon folded position
29 Dorsal spine fairing
30 UHF aerial
31 Tailcone frame and stringer construction
32 Magnetic compass remote transmitters
33 MAD detector housing and reeling unit
34 Tail rotor control cables
35 HF aerial cable
36 MAD unit fixed pylon
37 Ventral data link antenna housing
38 Lower UHF/TACAN aerial
39 Fuel jettison
40 Anti-collision light
41 Tie-down shackle
42 Tailcone joint frame
43 Air system heat-exchanger exhaust
44 Engine exhaust shroud
45 Emergency locator aerial
46 Engine fire suppression bottles
47 IFF aerial
48 Port side auxiliary power unit (APU)
49 Oil cooler exhaust grille
50 Starboard side air-conditioning plant
51 Engine exhaust pipe
52 HF radio equipment bay
53 Sliding cabin door rail
54 Aft AN/ALQ 142 ESM aerial fairing, port and starboard
55 Tailwheel leg strut
56 Fireproof fuel tanks, port and starboard, total capacity 592 US gal. (1368 litres)
57 Starboard stores pylon
58 Castoring twin tailwheels
59 Torpedo parachute housing
60 Mk 46 lightweight torpedo
61 Cabin rear bulkhead
62 Passenger seat
63 Honeycomb cabin floor panelling
64 Sliding cabin door
65 Recovery Assist, Secure and Traversing (RAST) aircraft haul-down fitting
66 Ventral cargo hook, 6000-lb (2722 kg) capacity
67 Floor beam construction

68 Spring-loaded door segment in way of stores pylon
69 Pull-out emergency exit window panel
70 Pneumatic sonobuoy launch rack (125 sonobuoys)
71 Rescue hoist/winch
72 General Electric T700-GE-401 turboshaft engine (1690 s.h.p.)
73 Engine accessory equipment gearbox
74 Intake particle separator air duct
75 Engine bay firewall
76 Oil cooler fan

77 Rotor brake unit
78 Engine intake ducts
79 Maintenance step
80 Engine drive shafts
81 Bevel drive gearboxes
82 Central main reduction gearbox
83 Rotor control swash plate
84 Rotor mast
85 Blade pitch control rods
86 Bi-filar vibration absorber
87 Rotor head fairing
88 Main rotor head (elastomeric, non-lubricated, bearings)
89 Blade pitch control horn

90 Lead-lag damper
91 Individual blade folding joints, electrically actuated
92 Blade spar crack detectors
93 Blade root attachment joints
94 Main rotor composite blades
95 Port engine intake
96 Control equipment sliding access cover
97 Engine-driven accessory gearboxes
98 Hydraulic pump
99 Flight control servo units
100 Flight control hydro-mechanical mixer unit
101 Cabin roof panelling

102 Radar operator's seat
103 AN/APS 124 radar console
104 Tie-down shackle
105 Gearbox and engine mounting main frames
106 Maintenance steps
107 Main undercarriage leg mounting
108 Shock absorber leg strut
109 Starboard mainwheel
110 Pivoted axle beam
111 Starboard navigation light
112 Cockpit step/main axle fairing
113 Forward cabin access panel
114 Collective and cyclic pitch control rods

115 Sliding fairing guide rails
116 Cooling air grille
117 Main rotor blade glass-fibre skins
118 Honeycomb trailing-edge panel
119 Titanium tube blade spar
120 Rotor blade drooped leading-edge
121 Leading-edge anti-erosion sheathing
122 Fixed trailing-edge tab
123 Cockpit eyebrow window
124 Rear view mirrors
125 Overhead engine throttle and fuel cock control levers
126 Circuit breaker panel

127 Pilot's seat
128 Safety harness
129 Crash-resistant seat mounting
130 Pull-out emergency exit window panel
131 Flight-deck floor level
132 Cockpit door
133 Boarding step
134 AN/APS 124 search radar antenna
135 Ventral radome
136 Retractable landing/hovering lamp
137 Downward vision window
138 Yaw control rudder pedals
139 Cyclic pitch control column

140 Instrument panel
141 Centre instrument console
142 Stand-by compass
143 ATO/co-pilot's seat
144 Outside air temperature gauge
145 Instrument panel shroud
146 Air data probes
147 Windscreen panels
148 Windscreen wipers
149 Hinged nose compartment access panel
150 Pitot tubes
151 Avionics equipment bay
152 Forward data link antenna
153 Forward AN/ALQ 142 ESM aerial housings

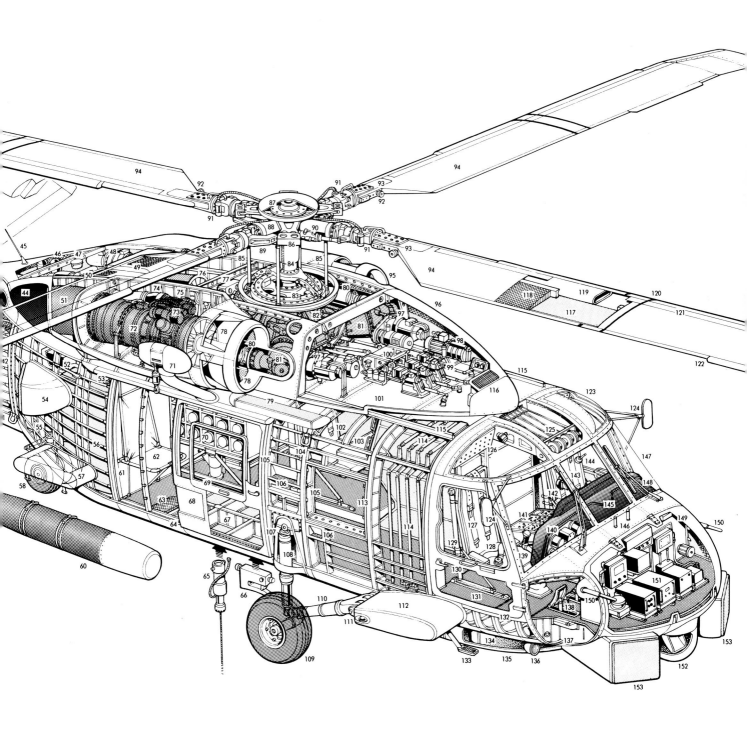

Operator on board a Royal Navy Sea King HAS.5 using the GEC Avionics AQS-902 to process and display data from dipping sonar and sonobuoys.

The Oliver Hazard Perry class guided-missile frigate USS McInerney, operated as a test ship for the Sikorsky Seahawk LAMPS III helicopter.

► GAF Searchmaster B basic coastal patrol version of the Nomad.

▼ French Navy Dassault-Breguet Gardians, operated since 1984 from their bases at Faaa, Tahiti and Tontouta.

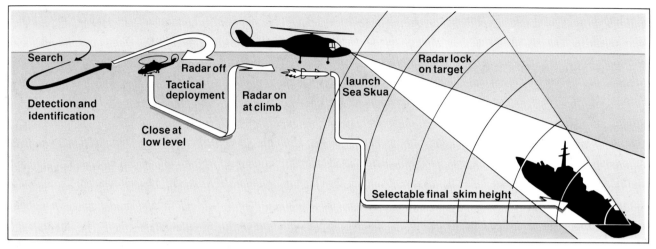

Sea Skua deployment and typical attack sequence (© British Aerospace).

a slight accident can change the signature). Nonetheless, the detectors can often identify individual submarines from their signatures. At worst, the MAD will detect the submarine and classify it in general terms for more detailed analysis by the sonobuoy system, and as a back-up to it. MAD is essentially a very short-range aid, the magnetic anomalies detected being very, very small; so the aircraft has to fly directly over the submarine, and usually it can only do that effectively if it is directed there by sonar information, either from its own sonobuoys or from sonar data derived by a ship in the group.

The ship/aircraft detection method is often used in association with helicopters. The mother ship will detect and classify a submarine and despatch its ASW helicopter, which can operate at a much greater distance than the range of the ship's weapons. Directed by radio link to the position determined by the ship's sonobuoy operator, the helicopter will confirm it is over the target using its MAD before launching weapons such as torpedoes or depth charges. While many helicopters are operated in this way, two of the latest helicopters for which this is a primary role are the US Navy's Sikorsky SH-60B Seahawk (built under the LAMPS III programme – Light Airborne Multi-Purpose System) and the Soviet Navy's new Kamov Ka-27 Helix-A.

Radar

The third sensor in a maritime patrol/ASW aircraft cannot 'see' under water but is nevertheless a vital piece of equipment. Radar is carried in the nose of many such aircraft, and its main purpose is to detect surface vessels at long ranges and in rough seas. It is, however, also tuned to be able to detect submarine periscopes at quite considerable ranges, perhaps up to 12 miles (20 km), so a submarine has to exercise extreme care when surfacing.

The radio transmissions of a submarine are also detectable by aircraft, giving the ASW crew another means of locating the vessel as well as providing it with vital information regarding the submarine's intentions. In certain instances it is possible for the aircraft to jam these transmissions, thereby denying the submarine contact with its base or the main fleet. Detection can also

be effected by 'sniffing' the atmosphere for traces of exhaust gases, but nuclear submarines emit no discernible exhaust and most submarines no longer have to use their periscopes, being able to rely on their own built-in passive sonars. Sonar and sonobuoys are therefore the prime detection agency in anti-submarine warfare.

All the sensors are controlled by the various crew members on the aircraft, the degree of analysis sophistication required by the strategists determining the size of platform and the number of crew employed. In larger ASW aircraft the data are then correlated on a central tactical console, so that the tactical controller has the full tactical picture. The aircraft's position can be stabilized at the centre of the screen and all the sonobuoys, with their detection tracks, the radar plot and any information provided by the magnetic anomaly detector and other sensors can all be displayed together; visual information, or data received from other aircraft or from base, can be displayed on the large screen to complete the picture.

Having located a submarine, in peacetime the data will be stored and then correlated back at base with all other data received so that individual submarines can be tracked throughout their voyages and any potential enemy's under-sea disposition is known continuously. In wartime, long-range maritime patrol aircraft are able to carry torpedoes, depth bombs, mines, missiles and so on in order to be able to launch a direct attack on a target, while smaller aircraft and helicopters can quickly call up support from the accompanying fleet or shore station if on-board weapons are inadequate, often handing over target information by data link to avoid any possible misunderstanding inherent in voice radio. Some helicopters, notably Soviet Navy Kamovs, are specifically equipped to provide air- or ship-launched cruise missiles with over-the-horizon target acquisition and mid-course guidance data.

Future trends

For the future, the cat-and-mouse game will continue unabated. While long-range aircraft will work autonomously, technology now permits those aircraft working within line of sight of a ship to communicate all their tactical information to it using a secure data link. This enables data from a number of co-operating aircraft to be correlated, along with the ship's own data, in one central operations room. The technique also has the advantage of reducing the number of crew members who need to be exposed to the dangers of such a vulnerable position, which saves lives and reduces costs and also enables the aircraft to be considerably smaller while performing the same task. This will be an important trend in the next generation of ASW aircraft.

Many navies are increasing their use of submarines; indeed recently New Zealand went as far as seriously considering changing over to an all-submarine fleet. As more submarines enter service, so there will be more ASW aircraft. Long-range ASW aircraft are very expensive and can be afforded by only a few nations, and in this sector of the market the Lockheed P-3 Orion will be unchallenged until at least the middle of the 1990s, when the US Navy starts to procure a replacement. Since there is little to be gained from specifying a new-technology airframe, it is quite possible that this replacement will be a development of the Orion itself, the emphasis being placed on even more capable electronics. The US Navy is also considering putting the S-3 Viking back into production after a gap of over 10 years, so as to be able to offer improved ASW protection to its fleet of aircraft carriers stationed throughout the world.

The French Atlantique 2 may be the next medium-sized aircraft to enter service, around the end of the 1980s, while in the very long term a consortium of British, German and US companies is studying a project which will be an ASW aircraft, civil and military transport and much more, all in one design.

At the smaller end of the market there has been a rush of new designs in the past ten years, typified by the British Aerospace Coastguarder, the GAF Searchmaster from Australia, the Fokker Maritime Enforcer from the Netherlands, and the Dassault-Breguet Gardian, aimed at providing affordable ASW and taking advantage of micro-electronics to give acceptable processing power in a small cabin. Even though many nations perceive a strong threat to security from under the waves, few have been able to justify the cost of purchasing and operating

Dassault-Breguet Gardian (France)
This is a maritime surveillance aircraft based upon the Mystère-Falcon 20. Five have been delivered to the French Navy, each with Thomson-CSF Varan search radar and other equipment. A small number of passengers or stretchers can still be accommodated if required.

Number in crew: Six
Engines: Two 5440 lb (2467 kg) thrust Garrett ATF3-6A-3C turbofans
Length: 56 ft 3 in. (17.15 m)
Wing span: 53 ft 6 in. (16.30 m)
Gross weight: 33,510 lb (15,200 kg)
Maximum speed: 438 mph (704 km/h)
Range: 2730 miles (4390 km)
Weapons: Attachment points under the fuselage and wings for equipment and weapons

Fokker F.27 Maritime and Maritime Enforcer (The Netherlands)
Fokker developed specialized maritime versions of its successful Friendship airliner to provide unarmed and armed surveillance aircraft for countries not requiring very expensive aircraft of the types built in the USA, Britain and France. The Maritime is unarmed and carries Litton AN/APS-504(V) 2 search radar (with 360° scan) in an underfuselage radome and other equipment, and is suited to coastal surveillance and SAR. Operators are Angola, the Netherlands, Nigeria, Peru, the Philippines, Spain and Thailand, the latter's three having the ability to carry weapons. The Maritime Enforcer is the standard armed model, carrying the same type of search radar plus radar detection equipment, MAD, sonobuoys and so on. This has the capability of ASW and ASV roles.

Data: Maritime
Number in crew: Six
Engines: Two 2140 s.h.p. Rolls-Royce Dart Mk 536-7R turboprops
Length: 77 ft 3½ in. (23.56 m)
Wing span: 95 ft 2 in. (29.00 m)
Gross weight: 45,000 lb (20,410 kg)
Cruising speed: 287 mph (463 km/h)
Range: 3107 miles (5000 km)
Weapons (Maritime Enforcer): Two or four torpedoes and/or depth bombs for ASW
Exocet or other air-to-surface anti-shipping missiles for ASV

Government Aircraft Factories Searchmaster (Australia)
The Searchmaster coastal patrol aircraft was developed from the 12–16 passenger GAF Nomad STOL transport. Two versions are operated, the Searchmaster B basic model carrying Bendix RDR1400 search radar with a forward-looking scanner in a nose radome, and the more sophisticated Searchmaster L with a Litton APS-504(V)2 radar and larger phased-array scanner that rotates in a radome located under the nose and other equipment. Indonesia, the Marshall Islands and Thailand all use Searchmasters.

Number in crew: Four in Searchmaster B, five in Searchmaster L
Engines: Two 420 s.h.p. Allison 250-B17C turboprops
Length: 47 ft 1 in. (14.35 m)
Wing span: 54 ft 2 in. (16.51 m)
Gross weight: 9100 lb (4127 kg) for Searchmaster L
Weapons: Up to 500 lb (227 kg) of weapons

even these relatively modest dedicated aircraft – the main way in which anti-submarine capability will enter nations' inventories will continue to be through the assisted sale of second-hand aircraft such as Orions and Il-38s to approved allies and friends, or the simple expedient of stationing a super-power's aircraft in another country.

While the airframes might not be new, the old adage about putting 'new wine in old bottles' will continue to keep airborne ASW as one of the leading defence technologies.

◁
USAF Boeing B-52G now serving in a maritime role, armed with Harpoon.

◁▼
Soviet Naval Aviation Tu-16 Badger-C anti-shipping aircraft with a wide nose radome and underwing pylons, engaged on a maritime mission.

▲
Super Etendard carrying the new ASMP (Air-Sol Moyenne Portée) missile with a warhead of 100–150 kiloton yield.

▶
Supersonic Tu-22M armed with a Kingfish missile, as photographed by an aircraft of the Swedish Air Force.

Tupolev Tu-16 and Xian H-6 (USSR/China)

The Tu-16 medium bomber entered Soviet Air Force service in the mid-1950s carrying free-fall conventional or nuclear bombs, but subsequent versions allowed for huge air-to-surface missiles or equipment for a variety of other roles including maritime and electronic reconnaissance, and ECM. Today the Tu-16 (NATO Badger) remains an important element of the Soviet Air Force and Naval Air Force, the latter having about 400 in service alone, of which only 75 are assigned as in-flight refuelling tankers. The bulk of the remainder are anti-shipping bombers in Badger-C, G and G modified forms and carrying Kipper or Kingfish, Kelt or Kingfish, or Kingfish air-to-surface missiles respectively. They are flown by the Baltic, Black Sea, Northern and Pacific Fleets. Of the remaining 80 Tu-16s or thereabouts, the Badger-D is the maritime and electronic reconnaissance model, Badger-E is a reconnaissance bomber, Badger-F is a reconnaissance bomber with an electronic intelligence pod, Badger-K is another electronic reconnaissance model, and Badgers H and J are ECM aircraft to confuse the enemy and escort strike forces. The Tu-16 has also been built in China as the Xian H-6 and reports have been made of H-6s carrying a Chinese anti-shipping missile known as the CAS-N-1, a Mach 2 missile with a range of 100 km and a 500 kg high-explosive warhead.

Data: Badger-G
Engines: Two 20,945 lb (9500 kg) thrust Mikulin
   RD-3M turbojets
Length: 114 ft 2 in. (34.80 m)
Wing span: 108 ft 0 in. (32.93 m)
Gross weight: 158,730 lb (72,000 kg)
Maximum speed: 616 mph (992 km/h)
Range: 2000 miles (3220 km) or more
Weapons: Up to seven 23 mm NR-23 cannon plus:
       Air-to-surface missiles as detailed above or
       up to 19,800 lb (9000 kg) of bombs

Anti-surface vessel

Surface vessels can be engaged by enemy warships and submarines, or challenged from the air and coastal defence systems. Although other chapters deal with air weaponry and the operation of carrier-based fighters and strike aircraft, it is worth mentioning here the operation of the large land-based aircraft whose role is that of anti-shipping.

Apart from the maritime patrol aircraft that can carry anti-shipping weapons and the many sea- and land-based fighter-bombers and helicopters able to launch similar missiles, both the USA and USSR have versions of their large land bombers now employed in anti-shipping roles.

The master of this role, both past and present, is the Soviet Union, which has produced specialized versions of its Tu-16 Badger, Tu-142 Bear, and supersonic Tu-22M

Tupolev Tu-22M (USSR)

Said in US circles to represent in wartime a threat to shipping every bit as great as Soviet submarines, the Soviet Naval Air Forces' fleet of Tu-22M supersonic reconnaissance-bombers operate over the Atlantic and Pacific. In 1985, a total of 105 were thought to be in Naval service, all of these being the Backfire-B version (NATO code name) and similar to one of three versions flown strategically by the Dalnaya Aviatsiya (Long Range Air Force) branch of the Soviet Air Force. The main weapon of the Backfire-B variable-geometry bombers in Naval operation is the Kingfish air-to-surface missile, which can have either a nuclear or conventional warhead. Other missiles can include Kitchen or the new AS-X-15 cruise missile with an 1860 mile (3000 km) range. There is no Western equivalent to this aircraft, which is still in production.

Data: Estimated
Number in crew: Four
Engines: Two turbofans, possibly 44,092 lb (20,000 kg)
   thrust Kuznetsov NK-144s
Length: 140 ft 0 in. (42.50 m)
Wing span: Spread 113 ft 0 in. (34.45 m)
         Swept 86 ft 0 in. (26.21 m)
Gross weight: 270,000 lb (122,500 kg)
Maximum speed: Nearly Mach 2
Mission radius: 3400 miles (5470 km)
Weapons: Two 23 mm cannon in the tail plus:
       Three Kitchen or Kingfish missiles normally
       Up to 26,450 lb (12,000 kg) of bombs

Backfire for anti-shipping purposes. Among weapons available to them are huge AS-2 Kipper and AS-6 Kingfish supersonic missiles, both with ranges of well over 100 nautical miles and the latter with alternative nuclear or high-explosive warheads. The 100 or so Backfires serving with Soviet Naval Aviation, and equipped with cruise missiles, have been said in the USA to pose perhaps a greater threat to the US fleets and forces necessary to resupply Europe than even the Soviet submarine force.

In contrast, USAF Boeing B-52G Stratofortresses not scheduled to be equipped with air-launched cruise missiles for a strategic role are replacing B-52Ds in the maritime role, equipped with the subsonic and conventionally armed Harpoon. Each B-52G carries 12 missiles and one squadron of 15 aircraft was planned for Atlantic operations and another for the Pacific. The possible strategic nuclear role of Soviet Naval Aviation bombers is partly offset by the US Navy's ability to operate its small carrier-based Intruders with nuclear weapons, although a substantial number of Intruders, like Orions and Nimrods, have been given Harpoon capability. The French Navy has also acquired an important strategic capability, having given carrier-based Super Etendards the capability of carrying the ASMP nuclear stand-off missile.

Sikorsky MH-53E Sea Dragon mine-countermeasures helicopter tows a magnetic-influence hydrofoil vehicle during trials.

A US Navy 'first'

The use of mines to deny an enemy access to harbours and ports is certainly nothing new, and has been well tried and tested as a means of providing an instant barrier to surface vessels and submarines alike. Maritime aircraft have the provision to carry mines, and have also been used since the Second World War for mine countermeasures. In 1970 the US Navy formed squadron HM-12 as its first *helicopter* mine-countermeasures unit, an important development in mine warfare. Each subsequently-received Sikorsky RH-53D helicopter had equipment deployed by a tow, designed to deal with mechanical, magnetic and acoustic mines. Since then Sikorsky has developed a much-improved airborne mine-

countermeasures (AMCM) helicopter as the MH-53E Sea Dragon, deliveries of which began in 1986. The Soviet Navy operates the Mil Mi-14 Haze-B for mine-countermeasures, which is not dissimilar to the usual Haze-A shore-based ASW version of the Mi-14, but it carries a fuselage strake and pod and does not have MAD.

In the never-ending battle to stay one jump ahead of the threat, the use of quick-response aircraft to deal effectively with sub-sea and sea surface forces can only increase. This may become even more evident to the West in years to come if the Soviet Navy deploys (as it is expected to) sea-skimming aircraft known as *ekranoplans*, wing-in-ground-effect (WIG) ships-cum-aircraft that could be fully armed and cruise at perhaps 280 m.p.h. (450 km/h) over extremely long ranges to disrupt sea lanes or reinforce at very short notice. Similar technology is also known to the West but little attempt has been made to exploit it.

Chapter 4
Tactics of an air force at sea

The aircraft carrier represents the most flexible application of air power possible with its ability to turn up anywhere in the world, combined, in the case of the present US super-carriers and forthcoming Soviet *Kremlin* class, with an awesome strike capability. The flip side of the coin is that they are prime targets for an enemy.

The huge US Navy carriers represent the top end of the scale, both numerically and in offensive capability, although they will be equalled in firepower by the Soviet *Kremlin* class vessels in the 1990s. Packed with the munitions and fuel to supply a carrier air group typically consisting of 86 aircraft, both fixed and rotary wing, they are vulnerable to attack by conventional weapons; far more so, in fact, than an airfield, which can soak up a fair amount of punishment without being rendered totally unusable by virtue of its sheer area. Furthermore a carrier is far more liable to be attacked with nuclear weapons than a land base, because a nuclear attack on a carrier task force at sea would obviously carry less risk of escalating the conflict. Therefore the aircraft carrier needs exceptionally effective defences. On the other hand, in order to carry out its allotted tasks, it needs a sizeable complement of strike aircraft, and these in turn need protection while they are performing their missions. Of the 86 aircraft carried on *Nimitz* class carriers, 16 are dedicated to anti-submarine operations and 34 to the strike role, while just two squadrons containing 12 aircraft each carry the burden of defending the carrier against air attack, as well as providing fighter cover for the strike squadrons. The 24 defensive aircraft are few for the task, but this is in part compensated by the remaining 12 'force multipliers', greatly increasing the effectiveness of the fighters. We shall examine these later insofar as they directly affect fighting operations. There are many variations on a theme, but the two main carrier fighter roles centre on air defence, in which the fighter operates as an interceptor, and escort, in which it assumes an air-superiority role. Both these roles have conflicting requirements, aerodynamically, tactically, and in weaponry. As a direct consequence of the first requirement, the design of the carrier fighter must be a compromise, whereas the other two allow a certain degree of optimization.

Fleet air defence

A cardinal principle of warfare is that offensive action can only be taken effectively from a secure base, and the function of fleet air defence is to provide that security. Defence may be seen in terms of a counter to a threat; the threat must be examined and the best solution sought in order to nullify it, within the constraints of 'state of the art' technology, achievable numerical strengths, and, unavoidably, available funding.

For many years, the air threat was provided by the 'iron' bomb, and, to a lesser extent, the air-launched torpedo. The attacking aircraft had to fly close to the target force before releasing its weapon, and this gave

Preceding pages
E-2C Hawkeye attached to USS America.

▶

F-8 Crusader on a side deck lift.

▲
Douglas F4D-1 Skyray, designed to carry four 20 mm cannon in the wings as an interceptor but with attachment points for bombs and rockets in a strike role. For the final period of service, a number of Skyrays were modified to carry Sidewinder air-to-air missiles.

◄
Rear port quarter view of USS Nimitz.

▷
F-4 Phantom IIs from Navy Squadron VF-96, then based on USS America.

orthodox radar-controlled fighter defences a good chance of intercepting in time. During the early 1950s the threat shifted to the fast high-flying jet bomber, which gave a far shorter reaction time to the defenders, who responded with interceptors optimized for high speed and a fast rate of climb. Typical of this breed was the Douglas F4D Skyray. The Skyray was armed with cannon which meant that it had to execute a perfect interception and get in close for a guns kill. The advent of workable homing missiles was the next step for the defences; this was quickly allied to supersonic fighters which possessed a decisive performance advantage over the fast jet bombers likely to be encountered. The most important of these was the F-8 Crusader, armed with two Sidewinder missiles and four 20 mm cannon. Although later variants had some adverse-weather capability, this was actually very

limited, and advances in avionics meant that a task force could be attacked around the clock in bad weather. The F-8 was essentially a clear-weather fighter and was unable to meet all projected threats.

In 1955, new fleet air defence requirements were laid down. The US Navy wanted a dedicated interceptor that could operate in adverse weather conditions, could stay on patrol for an extended period at a radius of 250 nautical miles from the carrier, was equipped with an advanced (for the day) detection and attack system, and was fast enough to catch the most modern jet bombers. This specification was responsible for the McDonnell Douglas F-4 Phantom II, which was capable of speeds in excess of Mach 2 and possessed an extended loiter time (although not both together), and had a good rate of climb. For detection it was fitted with the Westinghouse APQ-50

radar operated by a second crew member, and its armament initially consisted of four or six semi-active radar homing (SARH) Sparrow III missiles with a theoretical range of 22 nautical miles.

Defence procurement is a dynamic process and, having initiated the Phantom programme, the US Navy looked to the future. The next-generation threat appeared to be the fast jet bomber armed with what are now called cruise missiles, but at the time were known as stand-off bombs. These would be launched up to 200 nautical miles away from a target task force. Unless the launching aircraft could be intercepted before releasing its weapons, the missiles themselves would have to be intercepted, which compounded the problem considerably. The need was therefore to intercept the fast and high-flying bomber before its missiles were even launched. Excellent interceptor though the Phantom was, the patrols would have to be in the right place at the right time to stand any chance of achieving this. As this could not be relied on, a new solution was sought.

The initial proposal was to allow the weapons to do the work of interception rather than the fighter. The Bomarc surface-to-air missile was already being developed for this purpose by the USAF, and was for all practical purposes considered as a pilotless *ramjaeger*. For US Navy use, the AAM-N-10 Eagle was selected, which had a maximum speed of Mach 4 and a range of up to 110 nautical miles. To carry it, the Douglas F6D Missileer was selected, in essence a subsonic long-range patrol aircraft, able to carry six or eight Eagles and to remain on station for four hours or more. The Missileer was equipped with a Hughes long-range radar detection and missile control system, which provided in-flight guidance to the Eagles during the mid-course stage, terminal guidance over the final 10 nautical miles being by the missile's own active radar system.

The Missileer/Eagle combination was still-born. While it offered certain advantages, the disadvantages more than outweighed these. Once its weapons had been expended, the Missileer was defenceless, and it was far too slow to escape an enemy fighter. Weighed down by the heavy Eagles (over half a ton each), its time to operating altitude was poor; in consequence reinforcement was a slow process. It was also a single-role fighter that could not be used to escort the strike aircraft, and the limited space on board the aircraft carrier made it a luxury.

On the other hand, the US Navy could not afford to be without defences of this quality. The radar intended for the Missileer was developed further, while the technology from the Eagle was transferred from Bendix to Hughes, who combined it with their own AIM-47 Falcon to produce the AIM-54 Phoenix, matched to the Hughes detection and fire control system, AWG-9. At first, AWG-9 and Phoenix were intended to arm the swing-wing General Dynamics F-111B carrier fighter; when this project proved abortive, they were switched to the very successful F-14 Tomcat, which was designed by Grumman to be an uncompromising air-superiority fighter that was big enough to carry six Phoenix missiles.

Fleet air defence: the team

Enormously capable as the Tomcat is, the current threat is such that in the fleet air defence role it will rarely (if ever) have to fight while dependent entirely on its own resources. To be successful, any air attack on a US CBG would need to be carried out in force, with maximum use of feints, diversions, co-ordinated attacks and electronic

Grumman E-2C Hawkeye

1 Two-section rudder panels
2 Starboard outboard fin
3 Glass-fibre fin construction
4 Passive defence system antenna (PDS)
5 Rudder construction
6 Static discharger
7 Fin construction
8 Leading-edge de-icing
9 Wing-fold jury strut lock
10 Wing-folded position
11 Rudder jack
12 PDS receivers
13 Starboard inboard rudder sections
14 Starboard inboard glass-fibre fin
15 Port elevator construction
16 Port inboard fixed fin
17 Port outboard rudder sections
18 Rudder controls
19 Tailplane construction

20 Fuel jettison pipes
21 Rearward PDS antenna
22 Tailplane fixing
23 Rear fuselage construction
24 Tailskid jack
25 Arresting hook
26 Tailskid
27 Arresting hook jack
28 Lower PDS receiver and antenna
29 Rear pressure dome
30 Toilet
31 Rotodome rear mounting struts
32 Rotating radar scanner housing (rotodome)
33 Rotodome edge de-icing

34 UHF aerial array, AN/APS-138 set
35 Pivot bearing housing
36 IFF aerial array
37 Rotodome motor
38 Hydraulic lifting jack
39 Front mounting support frame

40 Radar transmission line
41 Fuselage frame construction
42 Toilet compartment doorway
43 Antenna coupler
44 Rear cabin window
45 Air controller's seat
46 Radar and instrument panels

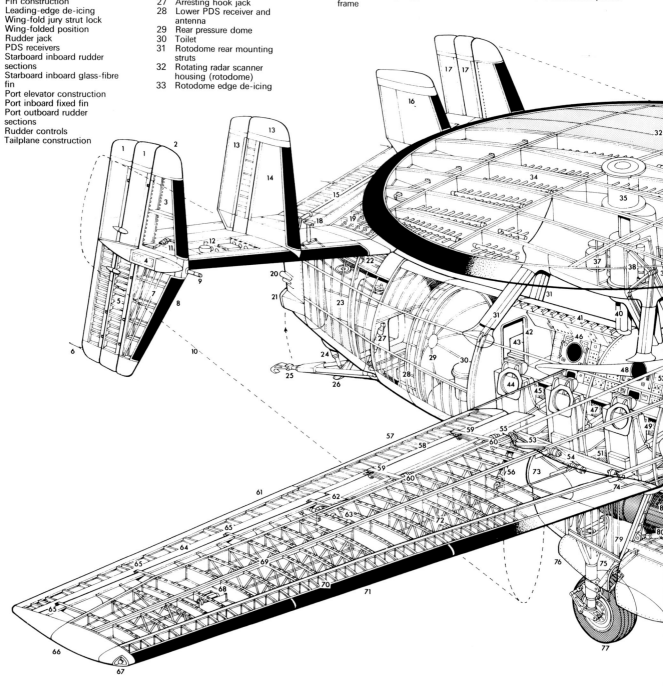

47 Combat information officer's seat
48 Combat information radar panel
49 Radar operator's seat
50 Radar panel and instruments
51 Swivelling seat mountings
52 Wing rear fixing
53 Wing-fold break-point
54 Spar locking mechanism
55 Wing-fold hinge
56 Wing-folding hydraulic jack
57 Starboard outboard flap
58 Flap construction
59 Flap guide rails
60 Flap drive motors and shaft
61 Starboard drooping aileron
62 Flap to drooping aileron connection
63 Aileron jack
64 Aileron construction
65 Aileron hinges
66 Starboard wing tip
67 Navigation light

68 Jury strut locking mechanism
69 Outer wing construction
70 Leading-edge construction
71 Leading-edge de-icing
72 Lattice rib construction
73 Engine exhaust pipe fairing
74 Front spar locking mechanism
75 Main undercarriage leg
76 Undercarriage leg door
77 Single mainwheel
78 Mainwheel door
79 Engine pylon construction
80 Engine mounting strut
81 Allison T56-A-425 engine (4910 e.h.p.)
82 Oil cooler
83 Oil cooler intake
84 Engine intake
85 Hamilton Standard four-bladed propeller
86 Gearbox drive shaft
87 Propeller mechanism
88 Cooling air intake
89 Engine-to-propeller gearbox
90 Oil tank, usable capacity 9.25 US gal. (35 litres) each nacelle
91 Bleed air supply duct

92 Vapour cycle air-conditioning plant
93 Wing front fixing
94 Computer bank
95 Wing centre rib joint
96 Inboard wing section fuel tank, capacity 912 US gal. (3452 litres) each wing
97 Lattice rib construction
98 Port inboard flap
99 Wing-fold hinge
100 Wing-fold joint line
101 Sloping hinge rib
102 Port outboard flap
103 Aileron jack
104 Port aileron
105 Port outer wing panel
106 Port wing tip
107 Navigation light
108 Leading-edge de-icing
109 Aileron control cable mechanism
110 Engine mounting strut attachment
111 Engine to propeller gearbox

112 Propeller spinner fairing
113 Hamilton Standard four-bladed propeller
114 Engine intake
115 Gearbox drive shaft
116 Port engine
117 Fuel system piping
118 Cooling air intake
119 Vapour cycle system radiator
120 Cooling air outlet duct
121 Radar processor
122 IFF processor
123 Radar transmission line
124 Rangefinder amplifier
125 Port side entry doorway
126 Equipment cooling air duct
127 Port side equipment racks
128 Starboard side radio and electronics racks
129 Radar duplexer
130 Electronics boxes
131 Forward fuselage frame construction
132 Lower electronics racks
133 Scrambler boxes
134 Navigation equipment
135 Cockpit air-conditioning duct
136 Cockpit doorway
137 Electrical system junction box
138 Air-conditioning diffuser
139 Signal equipment
140 Cockpit floor level
141 Co-pilot's seat
142 Parachute stowage
143 Pilot's seat
144 Headrest
145 Cockpit roof window
146 Cockpit roof construction
147 Instrument panel shroud
148 Windscreen wiper
149 Bulged cockpit side window
150 Instrument panel
151 Control column
152 Nose undercarriage strut
153 Nose undercarriage door
154 Rudder pedals
155 Nose construction
156 Pitot head
157 Sloping front bulkhead
158 Navigation code box
159 Nose electrical junction box
160 Rudder pedal linking mechanism
161 Windscreen heater unit
162 Nose undercarriage leg
163 Steering mechanism
164 Twin nosewheels
165 Catapult strop attachment arm
166 Nosewheel leg door
167 Nosewheel emergency air bottle
168 Nose PDS receivers
169 Oxygen tank
170 Landing lamp
171 Landing and taxi light window
172 Nose PDS antenna array
173 Nose aerial fairing

▲
Launch of an AIM-7F Sparrow.

▶
A US Navy technician checks the Hughes AN/AWG-9 weapons control system on a Tomcat fighter destined for the 'Screaming Eagles' Fighter Squadron 51 based on board USS Kitty Hawk.

▼
US fleet air defence—typical patrol cover.

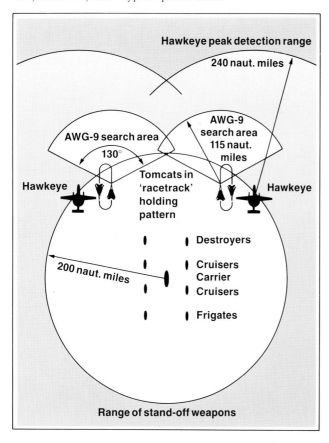

E-2C Hawkeye airborne early warning (AEW) aircraft. It is normal practice for two Tomcats to be on station together. They set up a racetrack holding pattern orientated towards the expected direction of threat, using their respective radars in the search mode. The shape of the radar scan is like a slice of pie cut at an angle of 130 degrees extending in front of the fighter. The racetrack pattern is adopted so that one fighter is always scanning in the direction of the expected threat. For long-range detection, two modes are available: pulse search, which has a nominal range of 62 nautical miles against a target with a five-square-metre radar reflecting area and little 'look-down' capability; and pulse-Doppler search, which gives a nominal detection range of 115 nautical miles against the same target, has good look-down capability (especially over the sea) but is blind to targets crossing the Tomcat's line of flight at or close to right-angles. Against targets with smaller radar reflecting areas, these ranges are very much reduced. There is always a possibility that a small fast target could penetrate the radar scan undetected and reach the blind area beneath the fighter. AWG-9 is very good but not infallible.

This is where the Grumman E-2C Hawkeye comes into its own. One Hawkeye shares a patrol station with each

countermeasures, using the most sophisticated weapons available. Without help, two squadrons of Tomcats, just 24 fighters, would be very hard pressed, and this is where the force-multipliers come in. The same would be true of carrier fighters of other nations.

The most common force-multiplier is the Grumman

▲▲

An EA–6B Prowler escorts Intruders on a strike mission exercise, jamming enemy radar emissions to render them electronically blind.

▲

The combat information centre officer, air control officer and radar operator in the ATDS compartment of a Hawkeye.

Grumman E-2 Hawkeye (USA)

The Hawkeye is probably the world's most successful AWACS aircraft, yet it was designed small enough to be compatible with US Navy aircraft carriers. It is also suitable for land operation, testifiable by its adoption by Egypt, Israel, Japan and Singapore, while US Navy aircraft have departed from their normal roles on occasions to monitor air traffic during launches of the Space Shuttle and have directed Coast Guard and Customs aircraft in anti-smuggling operations. In US Navy service, the Hawkeye's main role is surveillance and command, looking beyond the horizon or over difficult terrain to give warning to a carrier battle group of enemy activity. It can also direct friendly combat air patrols and co-ordinate friendly strike forces. The 'heart' of the Hawkeye is its Randtron AN/APA-171 above-fuselage rotodome radar and IFF antennae and General Electric AN/APS-138 advanced radar processing system with over-water and over-land detection capability, which can monitor more than 150,000 square miles of ocean surface and three million cubic miles of airspace. The radar system automatically locates and tracks airborne and surface targets, those of fighter size to a range of beyond 200 miles (322 km), larger targets (such as naval bombers) to the radar horizon, and anti-shipping cruise missiles to more than 125 miles (200 km), remaining effective even under enemy jamming operations. As well as the radar, other specialized equipment includes a passive detection system, listening to an enemy's signals from electronic emitters on board aircraft, ships and on land and thereby indicating threats before they come into radar range.

Data: E-2C Hawkeye
Number in crew: Five
Engines: Two 4910 e.h.p. Allison T56-A-425 turboprops
Length: 57 ft 6¾ in. (17.54 m)
Wing span: 80 ft 7 in. (24.56 m)
Gross weight: 51,933 lb (23,556 kg)
Maximum speed: 372 mph (598 km/h)
Endurance: More than 6 hours
Weapons: None

pair of Tomcats; it gives all-round radar coverage to a radius of 240 nautical miles and communicates with the pair of fighters over a two-way data link, which can transmit information straight on to the radar screen in the rear cockpit of each Tomcat.

Grumman EA-6B Prowler

1   Radome
2   APQ-92 radar antenna
3   Bulkhead
4   Rain removal nozzle
5   ALQ-126 receiver antenna fairing
6   Refuelling boom (detachable)
7   In-flight refuelling receptacle
8   Two-piece windscreens
9   Senior EWO's panoramic/video display consoles
10   Pilot's instrument panel shroud
11   Control column
12   Rudder pedals
13   Pitot static tubes (port and starboard)
14   Power brake
15   APQ-92 transmitter
16   Anti-collision beacon
17   'L'-band antenna
18   ALQ-92 (IFF) antenna
19   Taxi/landing light
20   Nosewheel leg fairing
21   Nosewheel leg
22   Tow link (landing position)
23   Tow link (launch position)
24   Dual nosewheel assembly
25   Nosewheel retraction jack
26   Nosewheel well door
27   Approach lights
28   Shock absorber link
29   APQ-92 high and low voltage
30   APQ-92 modulator
31   Cockpit floor level
32   Anti-skid control
33   Fuselage forward frames
34   Pilot's ejection seat
35   Senior Electronic Warfare Officer's (AN/ALQ-99 tactical jamming) ejection seat
36   Upward-hinged forward cockpit canopy
37   Canopy mechanism
38   Aft-cockpit port EWO's console
39   Handgrips
40   Security equipment
41   Splitter plate
42   Port engine intake
43   Intake frames
44   Aft-cockpit entry ladder
45   Electric hydraulic pump
46   Manual selector valves
47   Cockpit aft bulkhead
48   Third Electronic Warfare Officer's (ALQ-92 comms jamming) ejection seat
49   Second Electronic Warfare Officer's (AN/ALQ-99 tactical jamming) ejection seat
50   Canopy mechanism
51   Upward-hinged aft-cockpit canopy
52   Starboard outer ECM pod
53   Intake
54   Pod turbine power-source
55   ALQ-41/ALQ-100 starboard spear antenna
56   Leading-edge slats (deployed)
57   Starboard inner integral wing fuel cell
58   Starboard inner wing fence
59   Wing-fold cylinders
60   Hinge assembly
61   Wing-fold line
62   Starboard outer integral wing fuel cell
63   Fuel probe
64   Wing structure
65   Starboard outer wing fence
66   Starboard navigation light
67   Starboard formation light
68   Wing-tip speed-brakes (open)
69   Speed-brake actuating cylinder fairing

70   Fence
71   Wing-tip fuel dump outlet
72   Starboard single-slotted flap (outer section)
73   Starboard flaperons
74   Flaperon mechanism
75   Starboard single-slotted flap (inner section)
76   UHF/TACAN antenna
77   Directional control
78   Dorsal fairing frame
79   Computer power trim
80   Fuel lines
81   Control runs
82   Dorsal anti-collision beacon
83   Relay assembly group
84   Control linkage (bulkhead rear face)
85   Fuselage forward fuel cell
86   ALQ-126 receiver/transmitter
87   Hydraulic reservoir
88   Wing-root section front spar
89   Wing-root leading-edge spoiler
90   Engine bay frames
91   Port J52-P-408 turbojet (11,200 lb thrust)
92   Mainwheel door mechanism
93   Engine accessories
94   Mainwheel well door
95   Port mainwheel well
96   Transducer/accelerometer
97   Power distribution/transfer panels
98   Fuselage mid fuel cell
99   Roll trim actuator
100   Lateral actuator control
101   ARA-48 antenna
102   Vent lines
103   Fuselage aft fuel cell
104   Longitudinal control
105   Air-conditioning scoop
106   Fuel vent scoop
107   TACAN receiver
108   ALQ-92 air scoop
109   LOX (3)

110   Heat exchanger
111   Gyroscope assembly
112   Fuel control relay box
113   Adaptor-compensator compass
114   Arresting hook lift
115   Analogue to digital converter
116   Relay box/blanking unit
117   Control runs
118   Frequency and direction encoder
119   Fuel vent
120   Dorsal fillet

121   Starboard horizontal stabilizer
122   Multi-spar vertical stabilizer structure
123   Horizontal stabilizer actuator
124   Transmitter remote compass
125   Power divider
126   System Integration Receiver (SIR) antennae/receiver fairing

127   SIR antennae (Bands 4 and 7/8)
128   SIR receivers (Bands 4–9)
129   SIR antennae (Bands 4 and 5/6)
130   ALQ-41 transmit antennae
131   Attenuator
132   RF divider
133   Rudder upper hinge
134   Rudder (honeycomb structure)
135   Antenna (Band 1)
136   Antenna (Band 2)
137   Rudder lower hinge
138   Rear navigation light
139   ALQ-126 transmit antenna

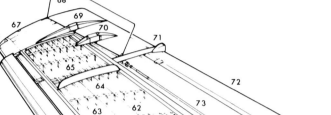

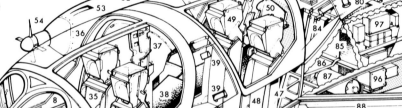

140 Fuel vent
141 Receiver antenna
142 Rudder actuator
143 Port horizontal stabilizer
    structure
144 Horizontal stabilizer pivot
145 Aft power supply
146 ALQ-41 transmitter
147 ALQ-41 receiver/transmitter
148 ALQ-100 receiver/transmitter
149 Chaff dispensers
150 UHF 'L'-band antenna
151 Arrestor hook
152 Extensible equipment
    platform (lowered)
153 APN-153 antenna
154 ALQ-41 power supply
155 ARC-105 radio receiver-
    transmitter
156 Power supply boxes
157 Port engine exhaust outlet
158 Wing/fuselage fairing
159 Ram air turbine (stowed)
160 Flaperon gearing actuator
161 Wing centre-section fuel cell
162 Port inner integral wing fuel
    cell
163 Port inner wing fence
164 Leading-edge slat structure

165 Wing-fold cylinder bays
166 Hinge assembly
167 Port flaperons
168 Flap actuator bays
169 Port single-slotted flap
    (outer section)
170 Wing-tip fuel dump outlet
171 Fence
172 Speed-brake actuating
    cylinder fairing
173 Wing-tip speed-brakes
    (open)
174 Port formation light
175 Port navigation light
176 Port outer wing fence
177 Leading-edge slats
178 Port outer integral wing fuel
    cell
179 Fuel probe

180 Port outer ALQ-99 high-
    power (tactical) noise-
    jamming systems pod
181 Port outer wing pylon
182 Port mainwheel
183 Mainwheel leg
184 Port inner wing pylon
185 Mainwheel retraction strut
186 ALQ-41/ALQ-100 (radar
    deception) port spear
    antenna
187 Port inner ALQ-99 systems
    pod
188 Garrett AiResearch four-
    bladed axial flow ram-air
    turbines
189 Ventral ALQ-99 high-power
    (tactical) noise-jamming
    systems pod

Grumman EA-6B Prowler (USA)

Based upon the airframe of the A-6 Intruder, the Prowler is an advanced electronic countermeasures aircraft and is the primary tactical jamming aircraft of the US Navy and Marine Corps. It can be both land based and carrier borne. Its sophisticated radar and communications jamming equipment allows it to protect a carrier battle group by screening it and jamming the sensors of long-range enemy bombers and cruise missiles, or the Prowler can protect friendly strike forces from enemy air defence systems. The 'heart' of the Prowler is its Eaton Corporation ALQ-99F tactical jamming system in five pods and with a total of ten transmitters, some or all of which can be carried according to the mission. The large pod at the top of the tail fin carries sensitive surveillance receivers which detect the emissions from enemy radars at long range. Detection, identification, direction finding and jammer-on-set sequence are performed automatically or with the assistance of the crew. The computer can be programmed before or during a mission using information from reconnaissance or ELINT (electronic intelligence) aircraft and reprogrammed if new threats are posed.

Number in crew: Four
Engines: Two 11,200 lb (5080 kg) thrust Pratt & Whitney J52-P-408 turbojets
Length: 59 ft 10 in. (18.24 m)
Wing span: 53 ft 0 in. (16.15 m)
Gross weight: 65,000 lb (29,483 kg)
Maximum speed: 651 mph (1048 km/h)
Mission range with auxiliary fuel: 2399 miles (3861 km)
Weapons: None

Crewmen on board USS Constellation *begin work on the surveillance receivers of a Prowler, used for the long-range detection of radars and carried in the large pod on top of the tailfin.*

The two-way data link is the next force-multiplier. Not only does it connect the Tomcats and the Hawkeyes, but it also communicates directly (and simultaneously) with fighter control on-board ship, thus helping to give the controller a comprehensive picture of the all-round tactical situation, and also the Grumman EA-6 Prowler electronic warfare aircraft.

The Prowler is also part of the team defending the carrier force. An attack has either to evade or overwhelm the defending Tomcats; it also needs to make radar contact with the surface vessels before it can launch its missiles. The primary function of the Prowler is to jam or spoof enemy radar emissions and render them electronically blind. To succeed in this even for a short while is valuable, as the attacking force will be unable to use the maximum range of their missiles and will be forced to close in to the carrier force to the point where the Tomcats can intercept them before the missiles are launched. A further function of the Prowler is to help overcome enemy jamming. Active electronic jamming involves an emission, the direction of which can often be detected. Working with the Hawkeyes, using the secure data link, this can boil down to a relatively simple triangulation problem to establish the positions of the enemy emitters, which in turn allows the Tomcats to take positive offensive action.

A fighter is often considered to be a single entity. In the context of the Tomcat, a great deal of teamwork is necessary between the front and rear crew members. In combat, the pilot flies and fights the aircraft, and has charge of the short-range weapons – Sidewinder missiles and cannon. But the Naval Flight Officer (NFO) is hardly superfluous. Apart from operating and monitoring the AWG-9 radar, it is his responsibility to decide on the tactics to be used and to set up the initial intercept. His decisions are based on the information shown on his target display indicator. This can be set either with the Tomcat represented in the centre of the display, which gives a check on whether anyone is creeping up behind with hostile intent, or with the Tomcat at the bottom of the screen, thus giving much more information on what is

Hawkeye searches for enemy aircraft and surface radar emissions, and directs attacking Intruders

Enemy missile sites

Attacking Intruders

Enemy missile range

Prowlers pinpoint enemy radars and jam the enemy air defence system

US Navy carrier-borne aircraft work together to defeat an enemy missile site (© Grumman Corporation).

KA-6D tanker, one of 62 converted from A-6A Intruders.

out in front. The AWG-9 can track up to 24 targets simultaneously while displaying as many as 12; data link can project a further eight on to the screen. With all this and two radios to handle, an NFO is likely to have his hands full.

As mentioned earlier, one of the prime assets of the Missileer was its long time on station. The Tomcat cannot hope to match this from its own resources, but an essential member of the carrier team is the KA-6 tanker; this provides in-flight refuelling and can be used to extend the interceptor's time on station, as well as giving essential fuel to thirsty fighters after combat, when the use of afterburners has run them short. The tanker completes the fleet air defence team: it increases the effectiveness of the fighters out of all proportion to their numbers.

Fleet air defence; the weapons
Before detailing the 'how to' section, it is necessary to consider the weaponry involved, as this has an important bearing on which tactics will be adopted. In turn, different weapons combinations provide optimum effectiveness against different threats. In combat, weapons selection will be decided according to the individual circumstances of that engagement.

Looking at the performance statistics of missiles in Chapter 5, the implication is that any target that approaches within range at a speed less than that of the missile is a 'dead duck'. In practice it does not work out like this. The speed, range and turn capability of the missile are all variable, the main factors being altitude, the speed and heading of the target, and the missile tracking limits. Brochure speeds and ranges for missiles are normally taken at high altitude, where the thin air minimizes drag and maximizes thrust. As a ball-park rule, missile range can be halved at 20,000 ft (6100 m) and halved again at sea level, although performance is

modified by the burn time of the sustainer motor if one is fitted. Maximum speed is less affected, but after motor burn-out the missile is coasting, losing speed because of the drag encountered. Attack ranges differ too. Once more as a rough guide, a high-speed head-on target can be attacked from three times as far away as is possible from astern. Beam (side-on) shots are between the two for range, but are more difficult to line up; furthermore, many proximity fuses are less effective from the beam.

The Tomcat carries a mix of four different weapons: the AIM-54 Phoenix, the AIM-7 Sparrow, the AIM-9 Sidewinder and the multi-barrelled M61 20 mm cannon. All are used in conjunction with the Tomcat's AWG-9 weapons control system, which, in its latest form, is equal to the world's best. Nothing being perfect, what are its weaknesses, also inherent in similar radars on others of the world's fighters?

Like any other radar-dependent system, it can be jammed or spoofed, although less so than most as it contains many clever electronic counter-counter-measures (ECCM) features. However, it is heavily dependent on pulse-Doppler radar, which is blind to a target crossing at a 90 degree angle off to a few degrees either way. Pulse radar can be used against these crossing targets but, unlike pulse-Doppler, it possesses little look-down capability, and therefore a low-flying crossing target is likely to remain undetected. This blind spot can be used by an enemy to evade detection.

Phoenix is the world's longest-range air-to-air missile. This immediately poses the problem of positive target identification. Only in cases where no doubt exists as to target identity can Phoenix be used; needless to say, some tactical situations preclude its use. On the other hand, its long range is coupled with the multi-shot ability of AWG-9. Six targets can be tracked simultaneously and a Phoenix launched at each. AWG-9 provides SARH on a time-sharing basis for mid-course guidance, with the terminal homing stage provided over the final 10 nautical miles by active radar in the nose of the missile. Active jamming is of course an emission, and, like a wasp at a picnic, Phoenix has what is described as a 'home on jam' facility.

Sparrow relies on SARH. The radar must be locked on to the target to illuminate it during the missile homing phase. This reduces the illuminating fighter's manoeuvring capability and makes it predictable for too long. Other weaknesses are due to the blind spots of the radar and the fact that Sparrow has little look-down capacity. The identification of targets is also a problem, as with all beyond-visual-distance missiles.

Sidewinder homes on the infra-red emissions (heat) of the target. Relatively short-ranged, it is essentially a visual distance weapon, and has the advantage that after launching it needs no further assistance from its parent fighter. While the latest models have an all-aspect capability, it is still best as a rear-quarter weapon. It can be spoofed by flares and confused by other heat sources, especially at low level. In thick cloud or heavy rain, its tracking ability drops dramatically.

All missiles have a minimum range within which they are unable to begin tracking. This area is covered by the gun, a reliable, instantly available, all-aspect weapon with an optimum range of about 1500 ft (460 m). It can be used at longer ranges, but this is the distance at which the average squadron pilot can be expected to hit anything.

Sparrows, Sidewinders and the M61 cannon are also among the weapons fitted to the latest US naval strike fighter, the McDonnell Douglas F/A-18A Hornet. The British-built Sea Harrier has provision for four AIM-9L Sidewinders in Royal Navy service, although Indian Navy Sea Harriers rely on the same type of French Matra Magic missiles as carried by carrier-based Super Etendards. The Magic is a similar type of missile to Sidewinder but has a shorter range and is highly manoeuvrable. At present the only fixed-wing aircraft element at sea within the Soviet Navy is the Yak-38; it is worth mentioning that it has provision for the AA-8 Aphid missile, which is remarkably like the Magic.

Fleet air defence: the tactics

Successful defence of the fleet depends on the outcome of the outer air battle. The carrier battle group will be spread widely over the ocean, with missile-armed cruisers and destroyers providing a defensive screen around the carrier. This is the 'last ditch' defence. The fighters operate well outside the missile engagement zone of the surface units, and, bearing in mind the extended range of current anti-ship missiles, they will attempt to intercept as far out as is practicable. The question is, what form will the attack(s) take? Much depends on the location of the battle group at the time. If it is far out in the ocean, the only feasible means of attack is by long-range bombers. Even then, variations on a theme are possible: a numerically strong swamping attack; a two- or even multi-pronged attack using a combination of high- and low-level formations, by day or by night, in fair weather or foul. If the battle group is closing (within, say, 400 nautical miles) to a hostile coast, which will be necessary if a strike is intended, any of these enemy attack combinations may be used, with the added complication of fighter escort. Alternatively, the enemy may decide to send out fighters in strength to whittle down the carrier fighters before launching its main attack. Let us examine

McDonnell Douglas F/A-18A Hornet (USA)
Designed as a lightweight strike fighter, capable of short field take-offs, catapult launching and arrested landings, the Hornet is entering service with the US Navy/Marine Corps to supersede three major aircraft, the F-4 Phantom II, A-4 Skyhawk and the A-7 Corsair II. The first unit to become operational with the Hornet, in early 1983, was Marine Fighter/Attack Squadron 314. A twin-finned, mid-wing fighter and attack aircraft, the USN/MC plan to acquire well over 1300 during many years of production, of which a reasonable number will be in TF/A-18A tandem two-seat trainer form. The cockpit layout for the pilot is particularly interesting, the use of three cathode ray tubes and an information control panel superseding much of the instrumentation of older aircraft. Carrier qualification trials with the Hornet were conducted in 1984, in which year the Hornet fleet surpassed the 100,000 flying hour mark.

Data: F/A-18A
Number in crew: One
Engines: Two 16,000 lb (7257 kg) thrust General
  Electric F404-GE-400 low-bypass turbofans
Length: 56 ft 0 in. (17.07 m)
Wing span: 40 ft 4¾ in. (12.31 m)
Gross weight: 49,200 lb (22,317 kg) in attack
  configuration
Maximum speed: Mach 1.8
Combat radius, internal fuel only: over 460 miles (740
  km) in fighter configuration
Weapons: One 20 mm M61 gun in nose plus:
    Up to six air-to-air missiles of
      Sidewinder/Sparrow types in fighter role
    Over 17,000 lb (7710 kg) of attack weapons,
      wingtip Sidewinders, auxiliary fuel or
      FLIR and laser spot tracker pods

two scenarios; a combined high/low attack by supersonic bombers coming from two different directions, and a fighter assault on fleet air defence fighters.

The primary requirement of successful air defence is early detection, and this should be supplied in the US Navy case by the Hawkeyes, operating on patrol lines which give overlapping coverage. The first threat to be detected, say, is a regiment of 18 supersonic bombers flying at high altitude – 40,000 ft (12,200 m) – and Mach 1.5. One minute's delay on the part of the defenders brings them 14 nautical miles closer to the carrier. They fly in very close formation, knowing that while they cannot hope to delay being detected by the Hawkeyes, their numbers, and therefore possibly their intentions, will be concealed for a while longer.

Inevitably the probing fingers of the Hawkeye radar detects the enemy force, whose reaction is to commence jamming. At the same time, the force splits, six bombers remaining at altitude, while the remaining dozen dive to just above sea level in the hope that this will enable them to evade detection longer. At sea level they are firmly subsonic, and trail the main force by a few miles, the distance between them slowly increasing.

▲
Tomcat carrying a mix of two Sidewinders (outboard), two Sparrows and four large Phoenix air-to-air missiles.

▶
Combat spread; except at very low level, formations always fly stepped down into the sun, so that it cannot form a blind spot to both fighters at once.

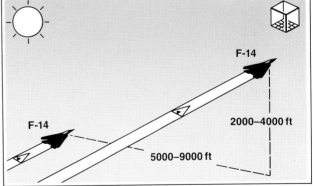

The jamming, although it fouls up the US radar detection network, confirms one thing: the intruders are hostile. The pair of Tomcats on station with the Hawkeye move into wide battle formation, or combat spread, engage zone 5 afterburner and accelerate towards the projected position of the threat on a lead collision course. This gives two advantages tactically. If their powerful AWG-9 can break through the jamming, the energy imparted to the Phoenix missiles at launching by virtue of the Tomcat's speed will increase its range, while, if all else fails, the interceptors will quickly reach a position from which a visual sighting can be made. A further possibility is that contact may be achieved by the infra-red sensor mounted beneath the nose of the Tomcat. This is normally slaved to the radar; it has inherently better angular resolution than radar and, being a passive sensor, its use cannot be detected. If necessary, it can even provide sufficient data to permit a Phoenix launch.

Meanwhile the attackers are in similar straits. Even though they are within missile range of their target, they cannot get the vital radar contact which will enable them to launch; the high-flying Prowlers are jamming them solid. Fleet air defence is a team effort, and each member plays his part. Meanwhile the net is closing around them as deck-launched interceptors are hurled off the catapults of the carrier to reinforce the defence. Other pairs of Tomcats working with other Hawkeyes remain on their patrol stations; while the risk of a pincer attack exists they must not be pulled out of position.

As the first pair of Tomcats closes to the high-level attackers, the electronic fog starts to lift and contact is gained. Warned by the pulse-Doppler emissions of AWG-9, the bombers turn away, gaining 90 degrees angle off and regaining a measure of invisibility, while at the same time diving to put themselves below the radar horizon where the pulse radar mode is less effective. But the

▲

Royal Navy Sea Harriers, each armed with four AIM-9L Sidewinders.

▶

Pulse Doppler radar-blind area. (Inset) Because its radar covers a so much greater volume of airspace than most other advanced fighters, the F-14 has a much better chance of finding a supersonic target and nearly two minutes extra in which to intercept it (© Grumman Corporation).

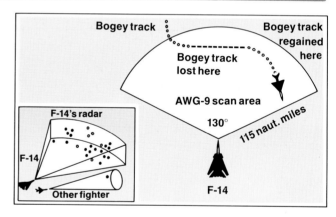

fighter pilots believe the bombers do not give up that easily; it must be a decoy move! The infra-red sensor is set to follow the high formation independently of the radar while AWG-9 is set to look down. It reveals the low unit, now many miles behind the high group. Data link transmits the information, which is reproduced in other Tomcats, the Hawkeyes and the ship-board command centre. The two Tomcats prepare their AIM-54s for launch. Target priority is assigned by one of three methods: from a controller with the overall tactical picture; by the AWG-9 fire control computer; or priorities can be assigned by the NFO in the rear seat of the fighter. In this instance, with two Tomcats in position to fire, target priority is assigned by the controller. This ensures that only one missile is launched at each target.

The Tomcat can carry up to six Phoenix, but they are heavy and bulky. If they are not expended, an arrested landing back on the carrier is a marginal affair due to the high all-up weight. In this case, with trouble expected, they are loaded with four Phoenix, two Sparrow and two Sidewinders. Eight AIM-54s do not go far with a dozen targets, and the controller makes the decision as to what happens next. If there is sufficient time they can continue to close for a head-on pass with Sparrow, although as this missile lacks the multi-shot capability of Phoenix, only one target can be attacked per fighter per pass; then again, time permitting, they can make a stern convergence for a Sidewinder attack, followed up with guns if

necessary. But with other deck-launched interceptors (DLIs) heading in at full speed, it is far more likely that the two Tomcats would be pulled out of the line of fire, possibly to refuel from a tanker, and to form a second line of defence.

Meanwhile, as the decoy bomber group is being intercepted by another pair of Tomcats, the second prong of the attack comes in, carefully timed to take advantage of the excitement and confusion generated by the first wave, at low level and high subsonic speed and from the far side of the battle group. Once again detection is followed by the invisible electronic battle, but this time the attackers try a different ploy. As they close towards the carrier, they launch several missiles. This move alters the priority for the defence. Even though the missiles were launched without the requisite radar lock-on being obtained, it is still possible that they may achieve a position from which the terminal homing phase might work. Shooting down the missiles, therefore, becomes equal priority to attacking the bombers. A hostile missile is a small radar target which can be acquired only at short range. During trials, a wave-skimming unaugmented drone has been downed by a Phoenix launched from a Tomcat 22 nautical miles away at a height of 10,000 ft

The pair of Tomcats on station with the Hawkeye before moving into combat spread and engaging zone 5 afterburner to meet the threat.

(3050 m). Having said that, the sea-skimming anti-ship missile is an extremely difficult target.

Tactically, fleet air defence against bomber attack is more a matter of winning the electronic battle, and of detecting in sufficient time and bringing enough firepower to bear on the threat, than of any clever manoeuvring. It is a team effort, and removing any part of the team reduces the effectiveness of the whole.

Directly enemy fighters enter the fray, the equation is radically altered. Bombers have little self-defence capability apart from speed, countermeasures, deception, and tailguns (if carried). Fighters can bite back.

Fighter pilot training includes the use of many basic fighter manoeuvres – how to outfight the opponent. While interesting, the majority of these are one-against-one situations, and as such they are peacetime training and bear little relation to war. The basic fighter element comprises a pair of aircraft; while larger units can also be used, they are built up of pairs. For the scenario here of a fighter attack on Tomcat fleet air defence patrols, it is therefore realistic to consider only the two-against-many event. It might conceivably be a two-against-two fight, but in war the golden rule is to expect the unexpected, and a two-against-two encounter can become 'multi-bogey' very quickly.

Students of air combat have identified five phases in an engagement: detection, closing, attack, manoeuvre and disengagement. To enlarge upon this, early detection followed by rapid closing should lead to a successful attack. The method of disengagement should be established before the closing phase is initiated. There is little point in launching even a successful attack if, at its conclusion, the attacker is left in a badly disadvantaged position, which is all too easy against fighters armed with modern missiles. So where does this leave manoeuvre? Manoeuvre is initiated by the aircraft under attack at some point during the closing phase or the attack itself. When this happens, the attacker must either break off and disengage, or manoeuvre himself to obtain a quick kill. In a multi-bogey situation the attacker cannot afford to stay in the manoeuvre phase for more than a few seconds, or he in turn will become the hunted. By contrast, the target fighter must manoeuvre as effectively as it can until the immediate danger is past.

Fighter combat is a subject for an entire book, so for the sake of simplicity AEW, ECM, and ECCM (airborne early warning, electronic countermeasures, and electronic counter-countermeasures) are omitted, to concentrate on the handling of a pair of fighters against a multiple threat. Here again the Tomcat is used as the example, although it applies in general terms to any fighter. Phoenix will rarely be used in this scenario; it is a very costly weapon and will be available only in limited quantities. If possible, it will be hoarded against the manned bomber attack. Therefore, consider the Tomcat as being armed with Sparrow, Sidewinder, and a gun, which is much closer to the possible armament of the world's other fighters.

The capabilities of AWG-9 make it likely, although not absolutely certain, that the Tomcat will detect its adversaries before it itself is detected, and this will confer the initiative. The first priority is to determine how many 'bogies' there are, along with their speed, course, altitude and formation. If one of the Tomcats is equipped with the Northrop Television Camera System (TCS), provided no clouds are in the way, the type of aircraft can also be determined; this helps a great deal, as the likely radar capability and weapons fit will then be known. The angle of the sun and cloud cover should not be overlooked either. All these factors help determine the tactics which will be employed.

The first decision to make is the choice of target(s). Is one of the enemy elements more vulnerable than the others (i.e. has it less immediate support, and does an attack on it offer an easy escape route)? The next decision is the method of attack, which will also determine the weapons and the formation to be used. Let us take some specific instances.

Six bogies are detected approaching at 20,000 ft (6100 m) in three pairs, each pair flying in wide battle formation about 5000 ft (1500 m) apart and with about two miles between each element. They have the sun immediately behind them, high in the sky. They are still some 50 nautical miles away, with a closing speed of about 1150 mph (1850 km/h). The Tomcats have less than three minutes to set up an attack. They select the pair on the left, dive to sea level and engage afterburner. Undetected closing is the next priority. Low-level flight will assist in this as they are masked against pulse radar by the sea clutter but, to stack the odds even more, they offset their course 30 degrees away from the bogies for half a minute before turning back on to a lead-collision course. Pulling up into a fast shallow climb, they select Sparrow and lock-on the radar. The in-range light flashes and the big missiles are launched, trailing their characteristic plumes of white smoke. The Tomcats follow, illuminating the targets until missile impact. One hit, one near miss! Pulling up into the vertical they commence a hard right turn across the sterns of the remaining bogies, who are now within visual range and can be identified. The instinctive reaction of the bogies has been to break into the direction of the attack, which puts them nearly 180 degrees out of phase with the Tomcats, who then have the option of disengaging by continuing on their present course, or using the vertical plane for a stern conversion before the enemy can turn around. The wisest course is disengagement, for even with a positional advantage long odds would be given for the success of two against five in a turning fight, which this would be. Also, more than a minute has passed since the radars were used for search; this is tiger country, and further bogies might materialize out of the blue at any second.

In another two-against-six scenario, the bogey elements are again in wide battle formation but, as the sun is to their left, each pair is stepped down into it to allow the best visual cross-cover. The elements trail each other, the first two at the same altitude and the third 10,000 ft (3050 m) lower. A front-quarter attack on the lead or second element will expose the Tomcats to retaliation from the rear element, which may well carry all-aspect missiles. The target selected is the rearmost element, while the Polish heart attack manoeuvre is chosen as giving the best chance of success. It is not just a glamorous name – it derives from the intercept geometry involved. Once again the Tomcats split for sea level, but this time they go very wide. Radar is left mainly on standby, using just the occasional scan to check that the bogies are still maintaining their course and formation. Moving fast, the Navy fighters pass down the flanks of the bogey formation and start to turn in just before they come abeam of the rear element. The weapon selected is the Sidewinder, and they each attack the aircraft farthest from them, crossing as they do so. If they are seen as they turn in – very likely, as the Tomcat is a large aeroplane – it will be by the nearest opponent, who will instinctively turn into what he perceives as an attack. If the Tomcats remain unseen, that is ideal; if they are spotted, the enemy fighter turning into the attack will perhaps set himself up for the attack coming in from his other side. Sidewinder does not need the assistance of the NFO in order to be launched; during the cross-turn the backseaters will have

Two-against-six combat event – first example.

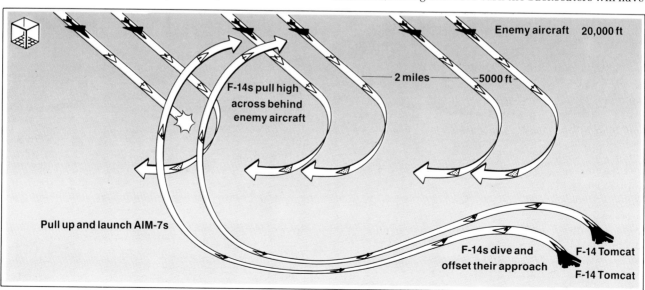

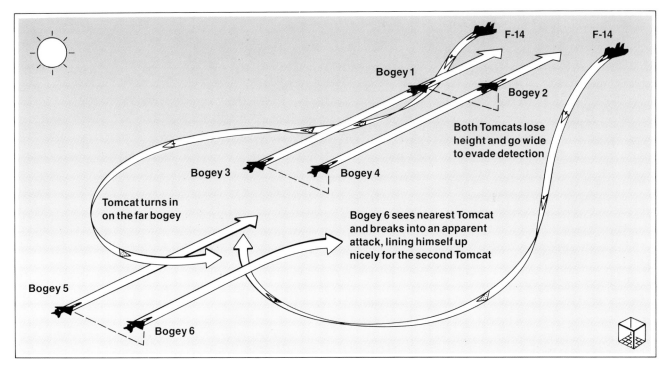

▲
Two against six—second example: the Polish-Heart attack.
►
Strike escort.

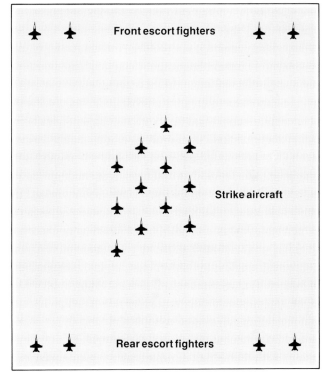

their 'heads out of the office' and be visually checking their own and each other's vulnerable belly and six o'clock positions. The attraction of this particular attack is that if the missile attack fails it can be followed up with guns, while the rest of the formation is nicely set up ahead.

What if either of these attacks goes wrong? The most obvious way for them to foul up is if they are detected on the initial run-in. Simply to turn away would be extremely dangerous against the jack-rabbit acceleration and medium-range missiles of a modern enemy fighter. To stay and fight at odds of three to one would be a better option but still not easy. The best means of evasion in both cases would be to blow through the enemy formation, passing close inboard of the nearest fighter to deny him room to turn, while launching missiles in the general direction of anyone bringing his nose to bear. It is not necessary for these missiles to be fired within parameters; the opponent will have no means of knowing whether they are or not, and the sight of a guided missile heading in one's direction is compelling. The usual effect is to make the pilot of the target aircraft forget any aggressive intent and to take the appropriate evasive action.

These two examples are typical of the out-numbered multi-bogey scenario, in that they exemplify a quick surprise strike, a rapid kill, avoidance of prolonged manoeuvre in the combat area, and a prompt departure. Many variations on the same theme are possible, depending on the relative positions at the start, the aircraft types involved and the weapons capabilities of both sides, and the numbers involved. The Tomcats do not have to use combat spread, but this formation lends

itself best to the teamwork that is necessary in all air fighting. While close combat is best avoided in a multi-bogey situation, tactical necessity may dictate otherwise. This tends to happen more often when escorting strikes over hostile territory than in fleet air defence, and we shall consider it in this context.

The escort mission

Escorting is the most difficult of all fighter missions. Historically it has proved impossible to protect friendly bombers completely, even with 100 or more fighters. Many types of escort are possible. The most effective is the fighter sweep, which precedes the strike force, ranging out across the most probable line of approach of the enemy fighters, on a seek-and-destroy mission. It may also be used to set up a CAP (combat air patrol) barrier. Second is the strike escort, a basic minimum of which is a flight of four in two elements, preceding the strike on each side, with a further flight bringing up the rear in similar fashion. Finally there is the reception committee, which is timed to meet the strike during its egress and to engage any pursuing fighters. The limiting factor is the carrier capacity; to provide anywhere near adequate protection is going to weaken the fleet air defence.

The strike complement of a *Nimitz* class carrier is 24 A-7 Corsair IIs split between two squadrons and a single squadron of 10 A-6 Intruders. The Intruder is an all-weather attack aircraft with a first-pass blind strike capability, and against heavily defended targets it would probably be reserved for night or bad-weather attacks. The Corsair is currently being phased out in favour of the F/A-18 Hornet. This puts a different complexion on escort work, as the Hornet is a true multi-role fighter. When operating in the attack role it is able to carry a Sidewinder on each wingtip, and even with a full bag of ordnance on board it is still far more manoeuvrable than the aircraft it replaces. Most pertinent is the fact that when its air-to-ground weapons have been expended, it can revert to the fighter role. Hornets on the return leg virtually constitute a fighter sweep! As a fighter it carries either six AIM-9s and two AIM-7s, or two AIM-9s and four AIM-7s. In close combat, it is a first-class performer.

It is reasonable to speculate that against a strong force, one Hornet squadron would supply the escort while the other flies the attack mission. At full strength this would provide eight Hornets for close escort and four in the fighter sweep/BARCAP. The reception committee could be drawn from the Tomcat squadrons. This would provide a formidable defence against fighter attack.

Close combat

Against determined fighter opposition, close combat would be unavoidable if the escorts were to keep the bandits away from the strike aircraft. Much of the fighting would be at visual ranges. In a confused multi-bogey air battle there would be little opportunity for the fighter pilot to use the formal manoeuvres and counter-manoeuvres that he will have learned in training. Every fighter has what is called a 'corner velocity', normally between 400 mph (650 km/h) and 520 mph (830 km/h), where its turn radius is smallest and its turn rate is greatest. This is where it is at its most manoeuvrable and it is often where victory and survival are decided. Ideally the fighter pilot should keep his speed above his aircraft's corner velocity so that he has a bit in hand for hard manoeuvring, which will bleed off speed rapidly. He should fly in a series of short, hard, unpredictable turns interspersed with straight-line accelerations to regain speed, taking shots of opportunity. There is a paradox in this situation, for while he needs to look ahead to score kills, he needs to look behind him to survive, and obviously he cannot do both at once. This type of air battle is very much the province of Sidewinder (or its foreign equivalent) and the gun; there is little opportunity to use a missile like Sparrow. Here and there chances will

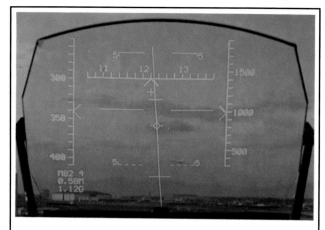

Head-Up Display (HUD)

Head-up display presents airspeed, heading and target information data on a transparent plate in the cockpit at the pilot's eye level. The HUD thus enables the pilot to receive relevant flight information without dipping his eyes from the forward view. One such HUD is produced by Hughes Aircraft and the bright symbology presented can be seen in the accompanying photograph. The plate is known as a combiner, the optical element in the HUD for displaying the information (Hughes Aircraft).

occur to manoeuvre into a favourable firing position, but the temptation to prolong the series of manoeuvres must be resisted unless his wingman is around to clear his tail. Even then the wingman may suddenly find himself engaged and unable to help. On the other hand, if the fight is multi-bogey for both sides, a certain amount of coincidental support will occur, and it must not be forgotten that the enemy is in the same straits. But confused dogfights notwithstanding, there are certain opening moves and end-game moves that can be used, just as in chess, although less precisely. These break down into offensive and defensive situations.

Offensive manoeuvring

The Polish heart attack mentioned earlier is a good example of offensive manoeuvring. There are other variations, mainly using an offensive split. Lateral changes of direction tend to affect the basic formation; a 45 degree change of tack puts the wingman into echelon while a 90 degree turn puts him into trail. Both of these positions can be utilized briefly, although as the trailing fighter has no cross-cover they should be used sparingly. Sometimes it is necessary to make visual identification; the hook serves very well for this. For the hook, the fighters use an offset low approach with the wingman wide and lower than the leader. The offset low approach has many advantages; it makes radar detection by the bogies more difficult by keeping below the horizon, and the offset serves to present the smallest (head-on) aspect to the bogies while approaching them from the front quarter, thus improving the chance of seeing them first. As they close, the wingman locks up the far bogey for a Sparrow shot while the leader closes to visual range. On identifying the bogies as hostile, the leader clears the

◄ *Pilot and naval flight officer under
the Tomcat's one-piece canopy.*

▼
A Hornet looses off a Sidewinder.

wingman to launch, then pulls high across the front of the bandits. This drags them for the wingman to make a stern convergence and a second firing opportunity, this time with Sidewinder.

Other offensive manoeuvres can be used once combat is joined; the high- and low-speed yoyos and the rollaway can be used to gain a firing position on a hard-turning defender, but in a confused mêlée only a few seconds can be spared before the attacker will be forced to break off and clear his own tail. These three manoeuvres are three-dimensional methods of defeating a two-dimensional manoeuvre which cannot be followed by normal means.

Defensive manoeuvres
Manoeuvre combat is normally initiated in order to defeat an attack. The first and most important is the break turn, a maximum-rate turn designed to defeat a missile attack or guns tracking. There is a slight difference between the two, although in each case the turn is made into the direction of the attack in order to gain as much angle-off

as possible. The missile break is generally made downwards, both to give the missile seeker head the most difficulty and to maintain aircraft energy, whereas the guns break involves a hard roll to get out of the plane of the attacking fighter. A guns attack is generally delivered at close range, and if there is a great disparity of speed between the defender and the attacker, the defender can use this to force an overshoot. Once he is satisfied that the attacker is overshooting, a quick reversal will turn the tables. If the attacker decides to stay in the fight, he will also reverse, and a scissors will result, the advantage being with the most manoeuvrable fighter. But a scissors bleeds off energy very quickly, and in a multi-bogey fight this is undesirable. It should only be attempted when there is a good chance of making a quick kill.

The sandwich is a very effective defence for a pair of fighters in combat spread. If attacked from astern and one of them can safely break outwards, when the attacker follows he sets himself up for the wingman to come in behind.

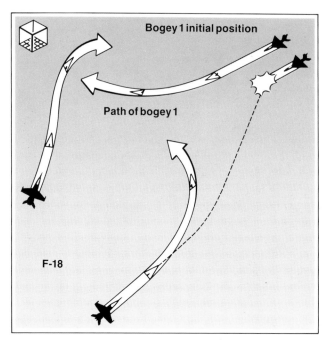

The hook: on visual contact the lead F-18 pulls up around the lead bogey to 'drag' it for a belly shot by his wingman. The wingman locks up the far bogey, and launches a Sparrow when the leader visually identifies the bogeys.

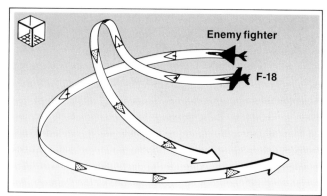

The high-speed yoyo uses the vertical plane to avoid overshooting a slower better-turning bandit.

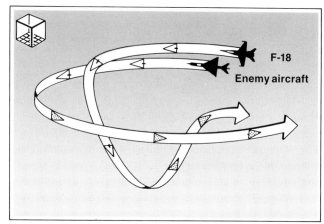

The low-speed yoyo: this is used to stay with a harder-turning bandit.

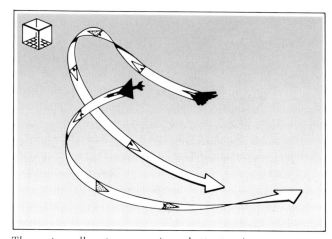

The vector roll: outmanoeuvring a better-turning opponent by rolling away from the direction of the turn.
▷

F/A-18A Hornets of Squadrons VFA-113 and VFA-25, now on their first extended deployment at sea on-board USS Constellation.

The defensive split is widely used by a pair when attacked from astern. One breaks high and the other low. Current fighter tactics are to follow the high man, who must be ready to bring the fight back down to regain support, while the instant the low man sees that he is not pursued, he should pitch up to lend assistance.

Disengagement from a multi-bogey fight is fraught with peril, yet the rate at which modern fighters consume fuel makes it of ever-increasing importance. Often it will involve a 180 degree change of heading, and this is most difficult to achieve while retaining mutual support. One method is to turn head-on towards an attacker and pass close by, thus denying him lateral turning room; by the time the attacker has turned around the defender will be long gone. The scissors can be broken off by the simple expedient of turning hard away, at a point when the opposing fighters are on diverging courses, then diving to pick up speed, while the split S can be used in many situations. This consists of a vertical dive, aileron-turning into the desired direction, then pulling out. Viewed from above, the split S enables a fighter to turn through 180 degrees with no lateral displacement. One of the best methods for a pair to use is the inward turnabout, wherein the pair turn towards each other through 180 degrees and resume their formation on the opposite course. This has the advantage of allowing the fighters to cover each other's tails during the turn.

The majority of the naval fighting power discussed has been from the viewpoint of the US super-carriers, because the US Navy has by far the greatest carrier forces afloat today. Thus, nations with smaller carriers and less sophisticated fighters are disadvantaged both in numbers and capability. The French Super Etendards in the attack role and elderly Crusader fighters, the Soviets' limited Yak-38 Forgers and the British Harriers simply

are not so effective except against limited opposition. The Sea Harrier, excellent fighter though it is, and the Forger, lack numerical strength and, most of all, lack the force-multipliers of the US Navy. Consider how much more effective the carriers Invincible and Hermes would have been in the South Atlantic in 1982 had they possessed airborne early warning and in-flight refuelling.

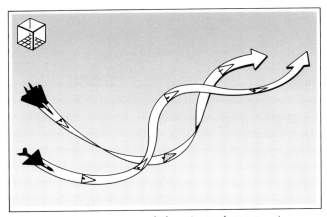

The scissors: not recommended against a better-turning opponent.

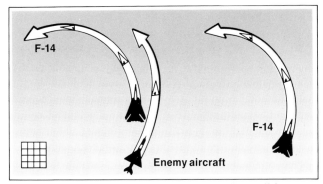

The sandwich: with combat spread an attacker will be caught like this if he does not disengage immediately he is spotted.

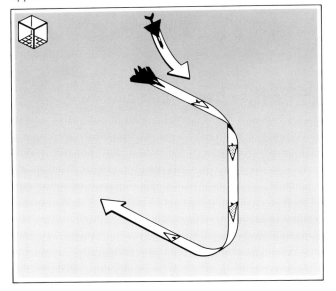

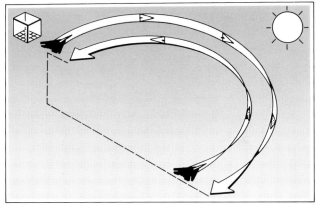

The inward turnabout: used to reverse the direction of flight while maintaining cross cover.

◄

Split S: a half-roll followed by a vertical dive gives a rapid increase in speed plus the ground masking effect against both radar and heat-seeking missiles.

Chapter 5
Naval air armament

Many types of armament carried by naval aircraft are identical to the weapons used by their land-based counterparts, but considerations peculiar to maritime warfare have produced additional demands. For example, a carrier battle group is such an inviting high-value target that hostile attacking aircraft must be intercepted before they can launch stand-off anti-ship missiles, hence naval fighters have often led in the development of long-range air-to-air guided weapons. Similarly, a heavily-armoured naval vessel, though easily detected, is difficult to destroy and therefore anti-ship guided weapons tend to have special types of guidance and warhead. The modern submarine, assuming it can be located, may only be vulnerable to a homing torpedo or a nuclear depth bomb. Here we deal firstly with multi-role armaments, then with weapons developed specifically for airborne, surface and sub-surface targets.

Multi-role armaments

Automatic weapons firing explosive ammunition, such as machine-guns and cannon, are probably still the most flexible type of armament. Although originally developed as a means to destroy bombers, the 30 mm DEFA cannon is believed to have been responsible for the destruction of the British frigate *Ardent* during the Falklands conflict in 1982, the weapon in this case being mounted on Aermacchi MB.339 light strike aircraft of the Argentine

Preceding pages
The British Aerospace Dynamics Sea Eagle is a 'fire and forget' anti-ship missile with an active radar homing head. It arms Royal Navy Sea Harriers, RAF Buccaneers and Indian Sea King helicopters.

▷ ▲
Close-up of the tail gunner's compartment of a Soviet Tu-142 Bear-D and the I-band tail warning radar above the turret. Enlargement of the tail blister showed the crewman waving a bottle of Pepsi!

▷ ▼
The US Navy's first operational surface-to-surface missile was the 575-mile range Chance Vought SSM-N-8 Regulus I, which eventually armed aircraft carriers, guided-missile vessels and submarines, and was also shore based. By 1957, 10 US carriers used Regulus, catapult-launched on a wheeled trolley.

Navy and firing armour piercing incendiary (API) rounds. The DEFA is one of several cannon families developed from the wartime German Mauser 213C series of single-barrel guns, which introduced the revolver principle derived from a cowboy's handgun. One of the latest revolver cannon is the British Aden 25, which fires the

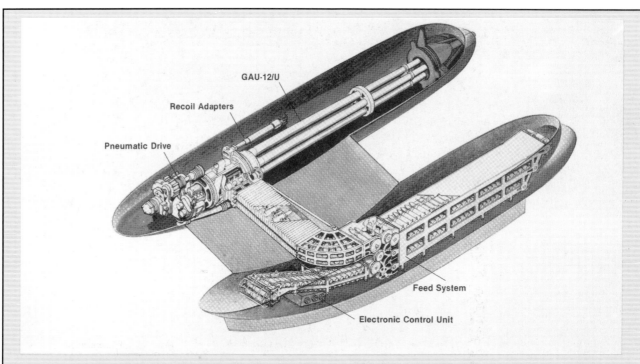

AV-8B Armament System

To provide gun armament to aircraft with no fixed weapons, several manufacturers produce podded gun systems which attach to the airframe and contain the gun itself and ammunition. For the new AV-8B Harrier II, General Electric has produced a system with two pods, called paks, which are bolted under the fuselage. Effective in air-to-air and air-to-ground roles, the port pak carries the 25 mm GAU-12/U five-barrel gun, blast deflector and pneumatic drive mechanism. The starboard pak contains a 300-round linear linkless ammunition feed system. A crossover fairing connects the paks and contains the ammunition chuting and drive shaft. The total weight of the Armament System is 1280 lb (581 kg) loaded (©General Electric).

◀

20 mm General Electric M61
Vulcan gun with six barrels.

▷

The array of bombs, rockets and
missiles that can be carried by the
Harrier II – mission loads can be:
(from front) four AIM-9L
Sidewinders; four LAU-10 and six
LAU-68 rocket launchers; six LAU-
61 and four LAU-68 rocket
launchers; six LAU-10 and four
LAU-61 rocket launchers; 10 Mk
77 520 lb bombs; 10 Mk 20 490 lb
bombs; 15 Mk 81 270 lb bombs; 16
Mk 82 531 lb bombs (and a triple-
ejector rack capable of carrying
three bombs); six Mk 83 985 lb
bombs; and four triple-ejector
racks. Other weapons can include
Maverick.

very powerful 25 mm Oerlikon KBB round at a rate of
1800 rounds per minute. Developed initially for the
British version of the Harrier II, this gun will almost
certainly be retro-fitted to the Royal Navy Sea Harrier.

After a brief flirtation with the revolver cannon, the
USA decided to adopt the Gatling rotating multi-barrel
concept, since this promised rates of fire that no
conventional gun could achieve. The principal result was
the General Electric 20 mm M61A1 Vulcan gun, with six
barrels firing 6000–7200 rounds/min. Although introduced in the late 1950s, the M61 series is still in service
with aircraft such as the Vought A-7E Corsair II,
Grumman F-14 Tomcat, and the McDonnell Douglas F/A-
18 Hornet. A more modern example is the 25 mm GAU-
12/U for the US Marine Corps' AV-8B, a five-barrel
weapon firing 3600 rounds/min. In smaller calibres, the
Gatling principle has been applied to 0.30 in. (7.62 mm)
machine-guns, producing the six-barrel GAU-2B/A
Minigun, which fires 6000 rounds/min. and is fitted to the
US Marine Corps' AH-1G and UH-1N helicopters, and the
US Navy's SH-3 helicopter. The 0.50 in. (12.7 mm)
GECAL-50 is a three-barrel 2000 rounds/min. gun
proposed for the Boeing–Bell JVX tilt-rotor series, which
may serve the Marines as the MV-22A Osprey medium
assault aircraft. The Soviet Union is known to have
developed a 12.7 mm Gatling gun for helicopter use, and
some reports suggest that a 30 mm Gatling may also be
available.

For the next generation, the most likely type of gun
appears to be a compromise between the single-barrel
revolver and the multi-barrel Gatling. Work in the USA is
proceeding on a twin-barrel cannon using 'telescoped'
ammunition, in which the projectile is sunk in the
propellant, thus reducing the overall length of the round.
The barrels are fixed, but use a combined operating

mechanism. This is basically a throw-back to the Hughes
20 mm Mk 11 gun of the 1950s, although the principle
originates from the German Gast machine-gun of World
War I. One of the standard Soviet aerial cannon is the
two-barrel 23 mm GSh-23, two of which are frequently
carried in podded form on the Yak-38 Forger.

Since the use of cannon by a fighter brought it within
the range of defensive fire from an enemy bomber,
unguided rockets were developed to produce a long-
range shotgun effect, a single hit supposedly resulting in
a quick and positive kill. These rockets doubled the reach
of the fighter, but their effectiveness has always been
regarded with suspicion. On the other hand, unguided
rockets are a cost-effective way of attacking a ship that is
neither heavily armoured nor well defended. British
naval aircraft use a 2 in. (51 mm) rocket, while the US
Navy employs a 2.75-in. (70 mm) projectile, and the
French produce the 68 mm SNEB for similar purposes.
Probably the largest weapon in this class is the US Navy
5-in. (127 mm) 'Zuni', which has largely been superseded
by guided weapons, although stocks are held against
possible future demands. Soviet aircraft employ a 57 mm
rocket. The Yak-38 often mounts two 16-tube pods of
these rockets, designation UB-16-57M, on the wing
pylons.

Unguided bombs are still an important form of
armament, although they take the aircraft very close to
the target, and their delivery accuracy is severely
degraded under wartime conditions. The standard US
Navy weapon is the Mk 82, weighing 531 lb (240 kg) in
low-drag general-purpose (LDGP) form and 570 lb (259
kg) in retarded 'Snakeye' form, which makes possible
safe and accurate delivery from a level release at low
altitude or in a shallow dive. The Royal Navy continues to
use the 1000 lb (454 kg) general purpose bomb in free-fall

form. During the Falklands conflict both Argentine and British bombs failed to explode in low-level attacks on shipping, since their fuses had been set for medium-level release. Naval aircraft also use cluster bombs to provide area coverage against personnel and lightly armoured targets, one example of this class of weapon being the Rockeye 490 lb (222 kg) bomb, containing 247 armour-piercing bomblets.

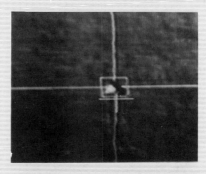

◁
*Forger-As on-board Novorossiysk, carrying underwing
GSh-23 gun and UB-16-57M rocket pods.*

Air-to-air

Turning to guided air-to-air weapons, these fall into three principal categories: short, medium, and long range. The most widely-used fairly short-range guided missile in the West is the AIM-9 Sidewinder, which achieved considerable success in the South Atlantic conflict of 1982, being responsible for 17 of the Sea Harrier's 23 air kills.

In an earlier war in which both land and naval air forces fought, Vietnam, the AIM-9G version of Sidewinder is reported to have had an average kill probability in the region of 50%, a remarkable figure for a lightweight missile. The more recent AIM-9L (as used in the Falklands conflict) is one of the few infra-red (IR) homing missiles that can make all-aspect attacks, since it employs a sufficiently long wavelength to detect the plume of hot gases that is produced by the engine, rather than having to look down the jetpipe for the really high temperatures in the turbine area.

However, most IR missiles can be decoyed by flares fired from the target, or by the sun, or even by clouds illuminated by the sun. Since guidance signals are traditionally generated by passing the target radiation through a rotating 'chopper' (the frequency variation and phase of the energy pulses then denoting the angular position of the target), missiles can also be thrown off course by a device generating IR pulses at a frequency related to the speed of the 'chopper'. A great deal of effort is therefore being directed towards making IR-homing missiles less susceptible to such countermeasures. Improved countermeasures-resistance is one of the features of the AIM-9M now being introduced into the US services, and of the Anglo-German advanced short-range air-to-air missile (ASRAAM).

It may be conjectured that ASRAAM employs an IR-imaging focal plane array consisting of several thousand sensors, giving far better target discrimination than a single cell. The ability to recognize a genuine target also makes it easier to use two-stage guidance (since the missile does not have to be locked on to the target prior to launch), and thus to exploit the full range potential of the weapon. Rapid progress in avionics is making possible a more 'intelligent' missile that can (for example) compute the speed of the target and reject anything (such as the sun) moving at a velocity outside a preset bracket. The Soviet Yak-38 is capable of carrying the AA-8 Aphid short-range IR-homing guided weapon, which reportedly has a launch weight of 121 lb (55 kg) and a maximum firing range of 3.8 nautical miles.

Developments in IR-homing weapons towards longer ranges now appear to be blurring the distinction between short- and medium-range missiles. It is probably significant that France is replacing both the short-range Magic 2 and medium-range Super-D (both developed by Matra) with a single weapon, the MICA (*Missile*

◁
*A Harrier of US Marine Corps Squadron VMA-542 fires a
5-in. Zuni rocket.*

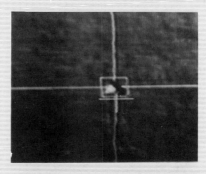
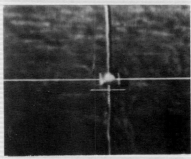
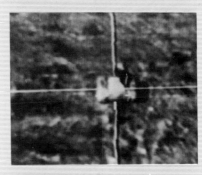
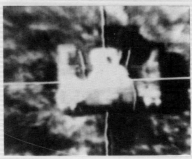

Imaging infra-red seeker

This remarkable sequence of photographs shows the imagery as seen by a Hughes Aircraft seeker as it closes on a tank target at night from a distance in excess of three nautical miles to impact. This seeker, developed by Hughes under a joint USAF/US Navy programme, was designed to be compatible with the GBU-15 glide weapon, Maverick and Walleye missiles.

The imaging infra-red missile seeker was developed to discriminate targets accurately through darkness, smoke or low-visibility haze. The IR seeker produces an image by sensing very small differences in infra-red heat radiated by objects in the missile's field of view (Hughes Aircraft).

Targets for training

Targets used to train gun and missile crews encompass many systems ranging from simple winched sleeves and model boats and aircraft, to highly sophisticated turbojet-powered high-speed targets. Among the airborne targets produced in various parts of the world, the British company Target Technology Ltd (TTL) produces the low-cost and recoverable BTT-3 Banshee, designed to simulate a missile threat. One user is the Sultanate of Oman, and a Banshee is seen here on board an Omani patrol boat on its catapult launcher. It is radio commanded and powered by a rear-mounted piston engine. Maximum speed is 200 mph (322 km/h) and it can remain airborne for up to 1 hr 15 min. (courtesy of Target Technology Ltd).

d'Interception et de Combat Aérien). While short-range missiles are being given a longer maximum firing range, there is also a move to reduce their minimum range to the order of 650 ft (200 m), so that they may at last replace the gun. One obvious way in which guided missiles can beat the gun is in 'off-boresight' engagements, making use of their lateral acceleration capability. However, once the pilot starts firing at targets outside the field of vision of his gunsight or head-up display, he needs some other means of aiming to ensure that the missile locks on to the intended aircraft. It thus seems highly likely that all future short-range missiles will be associated with a helmet-mounted sight.

Infra-red homing has always been the natural guidance system to use for a lightweight 'dogfight' missile, since such engagements are normally visual and since IR guidance produces a very small miss distance, which in turn makes possible a lightweight warhead and thus a small missile. However, IR homing does not work through cloud or rain; hence an all-weather missile demands radar homing. For maximum acquisition range and minimum missile cost, the natural solution is to rely on a semi-active missile that homes on to reflected energy from the fighter's radar. The most important Western wepon in this medium-range all-weather category is the US AIM-7 Sparrow series. Although fitted with a 66 lb (30 kg) warhead, a massive device in comparison with the 25 lb (11.4 kg) of the Sidewinder warhead, the AIM-7E is said to have achieved a kill probability of only 9% in Vietnam between 1965 and 1968. The AIM-7E was an interim weapon, and the AIM-7F with an 88 lb (40 kg) warhead and a maximum firing range extended to 24 nautical miles arrived in 1975, too late for that war. It is now being superseded by the AIM-7M with various

▶ *Two infra-red guided AIM-9L Sidewinders.*

improvements, including better electronic counter-measures (ECM) resistance and enhanced capability against low-flying aircraft.

One of the major restrictions on the use of the AIM-7 and any other semi-active radar-homing missile is that the launch aircraft has to continue towards the target until the weapon reaches it. In combat exercises it has been found that this allows the enemy time to launch a short-range missile before the AIM-7 arrives, hence both aircraft may be lost. To overcome this problem, the AIM-7 series is to be replaced by the Hughes AIM-120 advanced medium-range air-air missile (AMRAAM), a fully active weapon that will provide 'launch-and-leave' capability, and the ability to engage a series of up to six enemy aircraft in a ripple firing. This new weapon will have mid-course inertial guidance and active radar terminal homing, better electronic counter-countermeasures (ECCM), and improved performance against low-level targets. Some reports indicate that a later form of AMRAAM (AIM-120B) will have passive terminal homing, and will enter service around 1990.

At present Soviet naval aircraft do not appear to be capable of carrying medium-range missiles, but this deficiency is expected to be rectified with the introduction of the *Kremlin* class carriers. Reports suggest that the ship's possible fixed-wing complement of Su-27 Flanker fighters will be armed with AA-10 missiles, but carrier simulation trials have also included the MiG-29 Fulcrum (which is expected to use the same weapons).

One of the practical problems that restricts the use of beyond-visual range (BVR) missiles is identification of the target. In the Vietnam war it was often necessary when using the AIM-7 to either close in, identify the target, then drop back and fire, or to use a second aircraft to identify the target. Such procedures are cumbersome, and it is highly desirable that a more convenient method should be found. The F-14 is being fitted with the Northrop television camera set (TCS), a stabilized TV camera with a zoom lens, allowing a target to be

recognized and its external store configuration assessed at ten times the aircraft's visual acquisition range. The TCS can also automatically search for and track targets. A more ambitious approach is non-co-operative target recognition (NCTR) using a high-definition radar. Attempts at NCTR began some years ago with the Hughes AWG-9 system of the US Navy F-14A, and it may be assumed that they now extend to the APG-65 radar of the F/A-18.

The problem of target identification is even more severe in the case of long-range missiles, the principal example of which is the Hughes AIM-54 Phoenix. A few years ago the distinction between long- and medium-range missiles was that the former used multi-stage guidance. Now even short-range weapons are starting to use multi-stage guidance, and the distinguishing feature of long-range missiles appears to be their use of a high-altitude trajectory, to minimize the fuel needed to reach the target. In spite of such measures, long range is invariably associated with a weight penalty. According to public sources, the short-range AIM-9L weighs 191 lb (86.6 kg) and has a firing range of over 8.7 nautical miles, while the medium-range AIM-7F weighs 500 lb (227 kg) and can be fired 24 nautical miles, and the long-range AIM-54C weighs 1008 lb (458 kg) and has a range of over 108 nautical miles. However, this last figure is probably related to the use of a home-on-jam launch, and a more normal range against a non-jamming target may be in the region of 65 nautical miles.

Tests with the AIM-54 have nonetheless included some remarkable firings. The demonstrated range of 108 nautical miles refers to a head-on encounter between an F-14A flying at Mach 1.5 at 44,000 ft (13,400 m) and a BQM-34E target drone simulating a bomber with noise jammer, flying at Mach 1.5 and 50,000 ft (15,240 m). In this case the AIM-54 peaked at 103,500 ft (31,550 m). Other trial engagements include the ripple-firing of six AIM-54s against six drones flying at speeds ranging from Mach 0.6 to 1.1. In another test an AIM-54 was fired successfully from a range of 51 nautical miles against a Bomarc missile, simulating a bomber at Mach 2.8 and 72,000 ft (22,000 m).

Anti-shipping missiles

Air-to-air guided missiles played no significant part in the Second World War, but anti-shipping missiles were used successfully by the Germans in two forms, the Fritz-X and the Henschel Hs 293. The former was simply a radio-controlled glide-bomb, but the latter was rocket-powered, giving far more range. Between them they sank six ships (including the Italian battleship *Roma*) and damaged ten more. The Japanese Navy saw the potential of the missile somewhat later and produced what might be termed a man-guided missile, the Yokosuka MXY-7 Ohka (Cherry Blossom) suicide aircraft, which was dropped from Mitsubishi G4M2e Betty bombers. However, their numbers were not great and they were mainly defeated by naval fighters and proximity-fused anti-aircraft fire.

In the postwar era the surface vessel continued to represent a high-value, heavily-defended target, hence there was the same motivation to develop guided weapons to destroy ships from outside the range of their defensive fire. As in the case of the German weapons, initial developments employed command guidance, an operator (or the pilot) in the launch aircraft moving a small control stick which generated radio signals to the missile. Flares fitted to the rear of the missile made its position visible to the operator, who could thus judge the corrections necessary for its flight path. The most widely-used weapon in this class was the US Navy's Bullpup, which carried the warhead of a Mk 82 bomb in original form. However, command guidance had the disadvantage that the launch aircraft had to continue towards the ship, hence it was exposed to return fire, and weapon delivery accuracy thereby varied with the strength of defences. In the case of visual-based guidance, the system was limited to clear-weather daylight attacks, which restricted its operation. The majority of modern anti-ship missiles therefore use automatic radar command guidance or radar homing, either active (the missile having its own

▼

Distraction decoy: oncoming missile has choice of distraction decoys; also deals with simultaneous attack from different directions (© Plessey Aerospace).

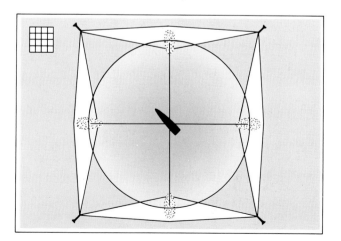

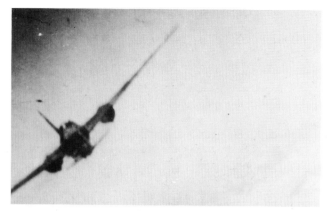

▲

Japan developed a specialized rocket-powered suicide aircraft as the single-seat Yokosuka MXY-7 Ohka, which was air-launched from a Mitsubishi G4M2e Betty bomber and covered some distance in a shallow glide before firing its motors. In the nose was packed 1200 kg of high-explosives. Many were intercepted before launch and these remarkable photographs show an Ohka-carrying G4M being intercepted in 1945.

▷ ▲

Bullpup carried by an A-4 Skyhawk.

▷ ▼

AN/APG-65 radar for the F/A-18 Hornet. The radome swings aside to allow technicians to slide the radar out for eye-level access to the entire system, including the five modular replaceable assemblies: the antenna, programmable signal processor, radar data processor, transmitter, and the receiver/exciter.

transmitter) or semi-active (homing on to reflected energy from the launch aircraft's radar).

Perhaps the best-known example is the Aérospatiale Exocet, which equips French Navy Super Etendards, Atlantiques and Super Frelon helicopters, and has also been sold to other nations. During the Falklands conflict the Type 42 destroyer HMS *Sheffield* and the container ship *Atlantic Conveyor* were both lost to Exocet strikes, the missiles being launched by Super Etendards of the Argentine Navy. It is generally believed that numerous ships have been sunk or seriously damaged in the Gulf War by Exocets fired from Iraqi aircraft. The air-launched AM.39 Exocet weighs 1438 lb (652 kg) and

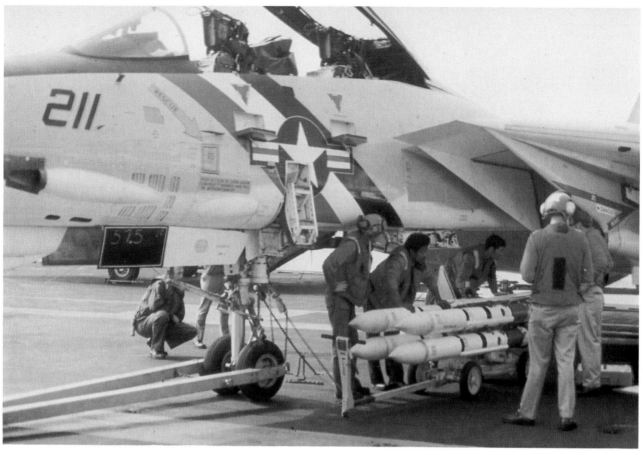

▲

AIM-54 Phoenix missiles are wheeled into position for mounting on a Tomcat by deckcrew red jacket armourers.

◄

The Italian Oto Melara Marte Mk 2 anti-ship missile for helicopters employs active radar homing, has 'fire and forget' capability which allows the aircraft to turn away after launch, and is insensible to ECM.

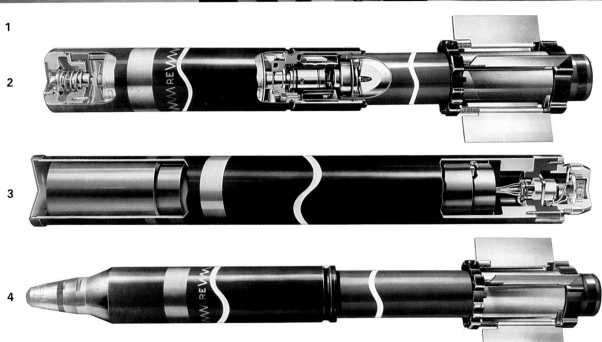

Shield anti-ship missile decoy system

One passive method of dealing with the threat posed by anti-ship missiles is the use of decoy systems, which attract the missile away from the intended target. The Plessey Shield system is in the forefront of this technology, which is based upon launchers for both chaff rockets and infra-red decoys, and a microprocessor-controlled command module. Designed to operate in conjunction with active electronic countermeasure systems, and close-in weapon systems, Shield can be fitted to warships of coastal patrol boat size and upwards. Plessey chaff and IR1 launchers are available in modules of 3 to 12 barrels in either crossed or parallel configurations (The Plessey Company Plc).

1 Parallel barrel launcher being loaded.

2 Plessey BBC (Broad Band Chaff) decoy rocket with folding fins and electronic fusing.

3 Plessey dual-frequency infra-red round which is programmed electronically.

4 Plessey BBC decoy rocket featuring folding fins and manual fusing.

Dassault-Breguet Super Etendard

1 Radome
2 Scanner housing
3 Flat plate radar scanner
4 Scanner tracking mechanism
5 Thomson-CSF/EMD Agave multi-mode radar equipment package
6 Refuelling probe housing
7 Retractable in-flight refuelling probe
8 Nav./attack avionics equipment
9 UHF aerial
10 Pitot head
11 Temperature probe
12 Refuelling probe retraction link and jack
13 Cockpit front pressure bulkhead
14 Instrument panel shroud
15 Windscreen panels
16 Head-up display
17 Control column
18 Rudder pedals
19 Cockpit section framing
20 Pressure floor level
21 Side console panel
22 Engine throttle lever
23 Radar hand controller
24 Nose undercarriage pivot fixing
25 Carrier deck approach lights
26 Nosewheel leg doors
27 Hydraulic steering jacks
28 Nosewheel forks
29 Nosewheel, aft retracting
30 Nose undercarriage leg strut
31 Rear breaker strut
32 Hydraulic retraction jack
33 Port engine air intake
34 Boundary layer splitter plate
35 Air-conditioning system ram air intake
36 Cockpit sloping rear pressure bulkhead
37 Boundary layer spill duct
38 Pilot's Hispano-built Martin-Baker SEMMB CM4A ejection seat
39 Starboard engine air intake
40 Ejection seat headrest
41 Face blind firing handle
42 Cockpit canopy cover, upward hingeing
43 Canopy hinge point
44 Canopy emergency release
45 Air-conditioning plant
46 Intake duct framing
47 Ventral cannon blast trough
48 Cannon barrel
49 Oxygen bottles (2)
50 Navigation and communications avionics equipment racks
51 Martin Pescador air-to-surface missile (Argentinian aircraft only)
52 Martin Pescador guidance pod
53 Starboard external fuel tank
54 Equipment bay dorsal access panels
55 Fuel system inverted-flight accumulator
56 Intake suction relief door
57 DEFA 30 mm cannon (2)
58 Ground power and intercom sockets
59 Ventral cannon pack access door

60 Ammunition magazine, 125 rounds per gun
61 Air system pre-cooler, avionics cooling air
62 Forward fuselage bag-type fuel tanks, total internal capacity 704 Imp. gal. (3200 litres)
63 Fuel tank access panels
64 Wing spar attachment main frames
65 Fuselage dorsal systems ducting
66 Avionics cooling air exit louvres
67 IFF aerial
68 Starboard wing integral fuel tank
69 Pylon attachment points
70 Matra 155 rocket launcher pod, 19 × 68 mm rockets
71 Leading-edge dog-tooth
72 Leading-edge flap control rod and links
73 Starboard leading-edge flap, lowered
74 Aileron hydraulic actuator
75 Wing-fold hydraulic jack
76 Outboard folding wing-tip panel
77 Strobe identification light
78 Starboard navigation light
79 Starboard wing-tip folded position
80 Fixed portion of trailing edge
81 Starboard aileron
82 Aileron hinge control
83 Aileron/spoiler interconnecting link
84 Spoiler hydraulic actuator
85 Starboard spoiler, open
86 Double-slotted Fowler-type flap, down position
87 Rear fuselage bag-type fuel tanks
88 Rudder control cables
89 Engine starter housing
90 Compressor intake
91 Forward engine mounting bulkhead
92 Accessory gearbox drive shaft

93 Gearbox-driven generators (2)
94 Engine accessory equipment
95 SNECMA Atar 8K-50 non-afterburning turbojet engine (11,025 lb thrust)
96 Engine bleed air duct to air-conditioning system
97 Rudder control cable quadrant
98 Fin spar attachment joint
99 Leading-edge access panel to control runs
100 All-moving tailplane pitch trim control electric motor
101 Tailplane root leading-edge aerodynamic notch
102 Elevator hydraulic actuator
103 Upper/lower fin segment joint
104 Tailplane sealing plate
105 Rudder hydraulic actuator
106 Tail-fin construction
107 Starboard all-moving tailplane
108 Forward radar warning antenna
109 VOR aerial (Argentinian aircraft only)
110 VHF aerial
111 Fin tip aerial fairing
112 Command telemetry aerial
113 Rudder
114 Rudder rib construction
115 Brake parachute housing, ground-based operations only
116 Tailcone parachute door
117 Tail navigation and anti-collision lights

118 Rear radar warning antenna
119 Port elevator
120 Elevator rib construction
121 Elevator damper
122 Port all-moving tailplane construction
123 Engine exhaust nozzle
124 Jet pipe
125 In-flight refuelling drogue, extended
126 Refuelling hose
127 Deck arrestor hook, lowered
128 Arrestor hook stowage fairing
129 Detachable tailcone frame and stringer construction
130 Rear fuselage break point, engine removal
131 Sloping fin spar attachment bulkhead
132 Engine bay heat shroud
133 Engine turbine section
134 Radar warning power amplifier
135 Fin spar and engine mounting bulkhead
136 Main engine mounting spigot

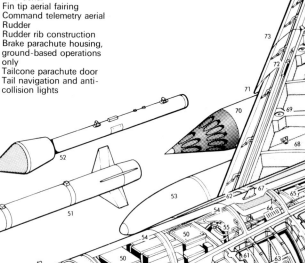

137 Aft avionics equipment bays, port and starboard
138 Port double-slotted Fowler-type flap
139 Flap rib construction
140 Flap shroud ribs
141 Inboard flap guide rail
142 Main undercarriage wheel bay
143 Main undercarriage leg pivot fixing
144 Flap hydraulic jack
145 Port spoiler
146 Spoiler hydraulic jack and control links
147 Outboard flap guide rail
148 Aileron rib construction
149 Port aileron
150 Port wing tip, folded position
151 Wing-tip panel construction
152 Wing-tip fairing

153 Port navigation light
154 Strobe identification light
155 Wing-fold hydraulic jack
156 Wing-fold hinge joints
157 Outboard leading-edge flap segment
158 Matra 550 Magic air-to-air missile
159 Missile launch rail
160 Matra 155 18 × 68 mm rocket pod
161 Aileron hydraulic actuator
162 Outboard pylon attachment joint
163 Outboard stores pylon
164 Leading-edge dog-tooth
165 Machined wing skin/stringer panel

166 Wing rib construction
167 Inboard pylon attachment joint
168 Inboard stores pylon
169 External fuel tank, 242 Imp. gal. (1100 litres)
170 Port mainwheel

171 Hydraulic multi-plate disc brake
172 Torque scissor links
173 Main undercarriage leg strut
174 Hydraulic retraction jack
175 Port wing integral fuel tank bays
176 Inboard leading-edge flap segment
177 Leading-edge flap rib construction
178 Ventral catapult strop hook
179 Wing-root bolted attachment joint
180 Leading-edge flap hydraulic jack
181 Extended chord wing root leading edge
182 Airbrake hydraulic jack
183 Ventral airbrake, port and starboard
184 Fuselage centreline pylon
185 In-flight refuelling 'Buddy' pack
186 AM 39 Exocet anti-ship missile

◁◁
Super Puma helicopter carrying AM39 Exocet.

◁
Royal Norwegian Air Force General Dynamics F-16A carrying a Penguin Mk 3 under its wing.

◁ ▼
A German MBB Kormoran missile skims towards its target using a radar terminal seeker; it can penetrate up to 90 mm of steel plate.

▶
A US Marine Corps A-4M Skyhawk from the China Lake Naval Weapons Center carries an AGM-65E Maverick with a laser seeker.

The Penguin Mk 3 has a programmed inertial mid-course guidance system and infra-red terminal homing. The missile utilizes any of three aircraft systems for targeting information: radar, head-up display or gunsight; it also has four phases of attack procedure: launch, mid-course, search and terminal. During mid-course the missile utilizes the self-contained inertial capability to proceed to the Waypoint determined by the pilot of the launch aircraft. On arrival, course and altitude are changed as programmed to intercept the target. In the search phase, the seeker is activated and one of several pre-selected search patterns is initiated as the missile enters the target area. When the seeker has acquired the target, it automatically locks on and guides the missile to impact.

◀
Harpoon being loaded on to an Orion.

▷
Soviet Badger-G modified with a wing-mounted Kingfish missile.

▼
Mines are among the effective anti-ship weapons. The British Aerospace Sea Urchin is a good example, able to be deployed by aircraft, surface ships or submarines in depths of 5–90 m. It is a programmable multi-influence ground mine and can be activated and triggered by the acoustic signature of a passing ship, a ship's magnetic influence or a change in water pressure from a ship's displacement. In addition, a target counting device provides an element of unpredictability.

carries a 353 lb (160 kg) warhead over a range of up to 37 nautical miles. It follows a sea-skimming flight path under the control of a radar altimeter, with inertial guidance followed by active radar homing in the terminal phase. One reported criticism made of Exocet is that it locks on to the largest radar target in the search area and thus can be decoyed by chaff.

The problem of chaff may in turn be countered by imaging infra-red (IIR) guidance, which in effect allows the missile to inspect a picture of the target, decide whether to attack it, and aim for the centre of the waterline. The IIR approach is employed in the US Navy's Hughes AGM-65F Maverick, and the Norwegian Kongsberg Penguin 3 missiles. Laser homing (to a spot on the ship illuminated by a laser designator) allows a specific part of the ship, such as the control centre, to be attacked. This guidance system is employed in the US Marine Corps' AGM-65E Maverick and the US Navy's Skipper 2, a rocket-powered laser-guided Mk 83 bomb.

For longer ranges, an air-breathing engine is required. The McDonnell Douglas AGM-84A Harpoon has a small turbojet and can achieve a range of more than 50 nautical miles when launched at altitude. It also has a pop-up terminal manoeuvre capability to attack the ship's deck, rather than the thicker side-armour. Harpoon is cleared for use on the US Navy's P-3 Orion and A-6 Intruder, with the S-3 Viking and F/A-18A as a second phase. It is also carried by maritime support B-52G Stratofortresses, and RAF Nimrods have similar provision. US Navy plans for an air-launched version of the Tomahawk (AGM-109L) cruise missile have currently been abandoned, but the air arm of the Soviet Navy employs a range of air-to-surface cruise missiles launched from heavy land-based bombers. The much smaller 6 nautical mile-range AS-7 Kerry tactical missile is reportedly used by the Yak-38 and employs simple radio command guidance. Weight is given as 2640 lb (1200 kg), with a 220 lb (100 kg) warhead. The Tu-16 Badger may carry one AS-2 Kipper with a range of 115 nautical miles or two AS-5 Kelts with a range

▼
Centroid decoy: 1. Chaff deployed close to ship. 2. Large chaff radar echoes remain as ship continues. Dispensed close to the surface of the sea, cloud radar echo is further enhanced due to multipath reflections between sea and chaff cloud (© Plessey Aerospace).

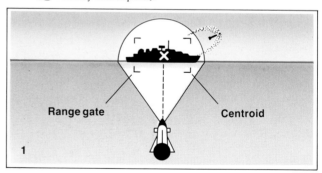

Range gate   **Centroid**

1

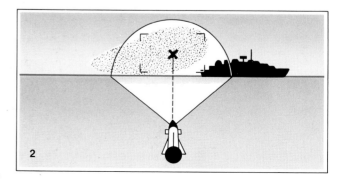

2

▼
Dump decoy: 1. Missile range gate locked on to the ship; chaff rocket launched and chaff cloud being dispersed. 2. ECM radar repeats and amplifies the ship's echo to the missile; chaff cloud developed. 3. ECM radar reduces the missile range gate and transfers it to the chaff cloud decoy. 4. ECM radar switched off; missile homes on chaff cloud echo. Ship no longer under threat and can take up challenge from other missiles (Ⓒ Plessey Aerospace).

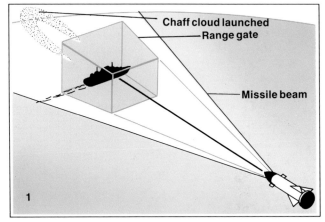

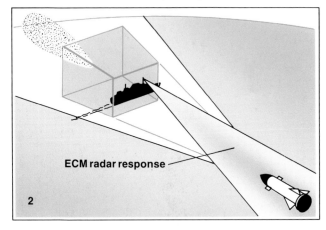

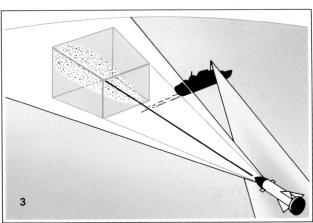

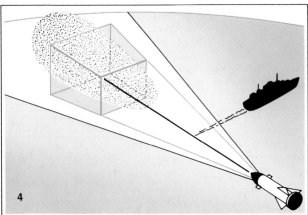

◁ ▲
Sea Skua lightweight anti-ship missiles on a Lynx.
◁ ▼
SA 365N testbed for the SA 365F/AS.15TT anti-ship version.
▲
*Soviet Backfire-B carrying an AS-4 Kitchen 37 ft-long
nuclear or high-explosive missile.*

of 85 nautical miles, or two Mach 3 AS-6 Kingfish with a
range of 120 nautical miles. The Tu-142 Bear can take one
350 nautical mile AS-3 Kangaroo on the centre-line or two
Mach 2 AS-4 Kitchens under the wings, the latter with a
range of 160 nautical miles. The Tu-22M Backfire is
associated with Kingfish and the AS-X-15 cruise missile,
the latter with a reported range of 1620 nautical miles.

At the opposite extreme, relatively small anti-ship
missiles are now available for use from helicopters that
are too light to carry Exocet or the British Aerospace
(BAe) Sea Eagle. The BAe Sea Skua was used to good
effect in the Falklands conflict from Royal Navy frigate-
borne Lynx helicopters. Eight Sea Skuas were fired and
all achieved direct hits. Whereas the British missile
employs semi-active radar guidance, the Aérospatiale
AS-15TT uses automatic command guidance, which gives
the lowest possible round cost. Before leaving the subject
of anti-ship missiles, mention must be made of the
Aérospatiale/MBB ANS (*Anti-Navire Supersonique*),
which will replace Exocet and will have ramjet
propulsion for long-range supersonic cruise.

Torpedoes for air launching

Helicopters of most navies are cleared to use lightweight
anti-submarine torpedoes fitted with either active or
passive sonar homing devices. Since the helicopter can
launch close to the submarine target, the torpedo is a
much smaller device than the heavyweight types used by
a submarine to destroy a battleship at long range. The
Royal Navy has generally followed the US Navy in this
class of weapon, beginning with the Mk 43 fitted to the
Wasp helicopter. This was followed by an Anglicized
version of the Mk 44, and then the Mk 46. However,
whereas the US Navy then adopted the Mk 46 Mod. 5
Neartip (near-term improvement), the Royal Navy
specified a major step forward in NASR 7511, which led
to the GEC Avionics Sting Ray, first deployed in 1982.
Designed for very high speeds (some 50% faster than the
latest submarines), comparatively long range and
operation at extreme depth, Sting Ray uses pump-jet
propulsion driven by an electric motor. The motor
receives power from a sea-water battery, which takes in
the electrolyte as the torpedo enters the water. In order to
pierce the stronger hulls of modern submarines, Sting
Ray has a shaped-charge warhead, similar to that of an
anti-tank missile. Other improvements include a very
large computer capacity, a frequency-modulated sonar,
and fast-acting controls to give a quick 'pull-out' when
Sting Ray is dropped in shallow water. Of course the
USSR, Sweden and France are among other countries
manufacturing air-launched torpedoes.

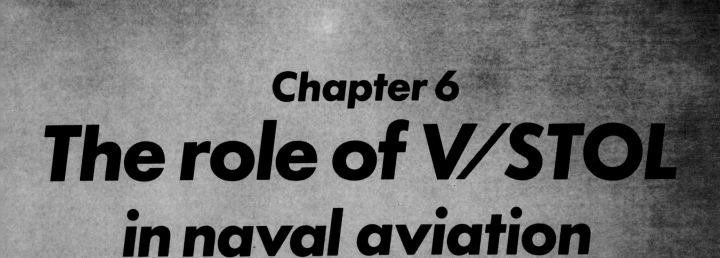

Chapter 6
The role of V/STOL
in naval aviation

The concept of an aircraft that can combine high-speed capability with V/STOL (vertical or short take-off and landing) became a reality in 1960 with the advent of the Hawker P.1127, powered by a single Bristol Aero Engines BE.53 vectored-thrust turbofan. With technical refinements and corporate restructuring, the aircraft became the British Aerospace Harrier and its engine the Rolls-Royce Pegasus, but the basic concept was unchanged. It was still a subsonic airframe with a centrally-mounted front-fan engine, the thrust of which could be turned (vectored) through an angle of 98.5° by moving an extra lever in the cockpit, thus rotating the four nozzles through which the engine efflux was discharged. In order to provide a thrust comparable with the weight of the aircraft, the Pegasus had been given a relatively high

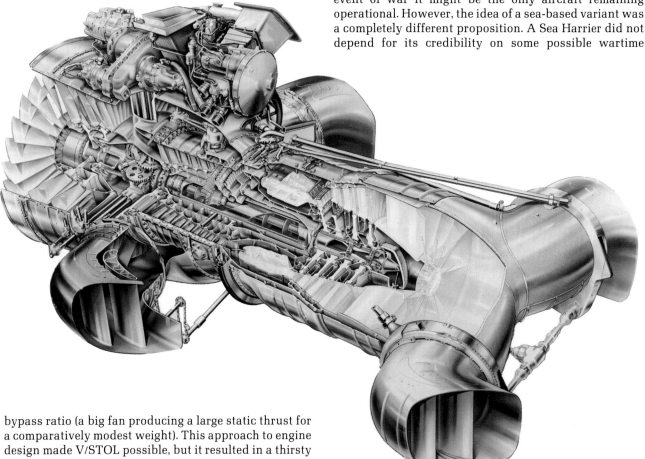

bypass ratio (a big fan producing a large static thrust for a comparatively modest weight). This approach to engine design made V/STOL possible, but it resulted in a thirsty engine for cruise flying.

After a fruitless diversion in the quest for supersonic speed in level flight, the Harrier finally entered service with Britain's RAF in 1969, and with the US Marine Corps (USMC) as the AV-8A in 1971. The USMC had special needs in the context of amphibious landings, which made V/STOL particularly attractive, but for the majority of air forces V/STOL was seen simply to represent additional expense and very limited in-flight performance. It was technically interesting, but the first priorities were supersonic capability and excellent warload–radius performance, which V/STOL aircraft could not provide. If conventional supersonic aircraft required airfield defence measures to keep them operating in war conditions, that was a problem operators understood. Although it might be attractive to

have a V/STOL element as insurance against the near total grounding of its air force through devastating attacks on airfields, few air forces seemed willing to consider this policy. Today many still view the option of attempting to land heavyweight supersonic conventional fighters (perhaps fitted with new STOL devices) between airfield craters as preferable to the subsonic but invulnerable VTOL aircraft, although in fairness supersonic performance is often an operational necessity and many lighter first-line supersonic fighters can operate from main roads and even grass strips.

The land-based Harrier therefore found it fairly difficult in the international marketplace. It did the same job as, say, a conventional A-4 Skyhawk but more expensively, and the only justification was that in the event of war it might be the only aircraft remaining operational. However, the idea of a sea-based variant was a completely different proposition. A Sea Harrier did not depend for its credibility on some possible wartime

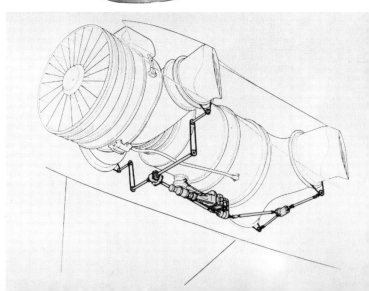

Preceding pages
Sea Harrier returning from a combat air patrol during the Falklands conflict, with the Royal Fleet Auxiliary replenishment ship Resource or Regent in the background.

The engine behind the Harrier's unique success, the Rolls-Royce Pegasus.

▼
Pegasus nozzle actuation system.

▶
Indian Navy Sea Harrier FRS.Mk 51s assigned to No 300 (White Tiger) Squadron to operate from Vikrant.

advantage: it offered the prospect of organic air power to navies that could not afford modern carriers to take conventional fixed-wing aircraft. Since it had no need for catapults and arrestor gear, the carrier for V/STOL aircraft could be a small, simply-equipped ship, possibly even a converted merchantman.

For navies with no fixed-wing aircraft, V/STOL had definite attractions. It made possible some measure of air defence outside the range of surface-to-air missiles, it offered over-the-horizon reconnaissance and, with modern anti-ship and other missiles, it promised good strike potential at several times the range of even a battleship's guns.

On further reflection, all kinds of secondary advantages came to the surface. Since it could take off vertically (admittedly at reduced weight), a V/STOL aircraft could provide an air defence capability even when its parent ship was in port, which might well be a third of the time. Since it could land vertically, the V/STOL aircraft could be recovered without turning the carrier into wind, which was a useful advantage when steaming through narrow straits, for example. Because they were free from the usual restrictions associated with catapults and arrestors, V/STOL aircraft could also be launched and recovered at a faster rate than conventional aircraft. During the recovery phase, more aircraft could be held on

US Marine Corps AV-8 armed with 30 mm Aden cannon and 2.75-in. rocket pods.

deck than in a conventional case, since vertical landing required little clearance from the next aircraft. The operational flexibility of V/STOL aircraft also meant that an enemy would have no certain knowledge of which ship was being used as a carrier at any given time. Any vessel with a helicopter platform was automatically transformed into a potential base for a V/STOL air defence fighter or strike aircraft.

Since the cost of a Sea Harrier is roughly the same as that of a Sea King helicopter, these operational advantages over conventional fixed-wing naval aircraft made it highly attractive to a number of navies. In addition, it appeared that the thrust-vectoring facility required to give it V/STOL could also be used in flight to enhance its manoeuvring effectiveness in close combat. If an enemy fighter closes in on its tail, the Sea Harrier can 'drop' (vector) its nozzles, which produces much the same effect as a large, invisible airbrake. The main result is that the attacker overshoots, while the Sea Harrier pitches nose-up and turns tighter, moving into a firing position behind its assailant. The use of thrust vectoring in forward flight (VIFF) is really a last-ditch manoeuvre but nevertheless an extremely useful feature, as is the ability to retain control at extremely low airspeeds due to the

reaction control system incorporated for V/STOL use.

From a naval viewpoint, V/STOL aircraft were certainly worthy of serious consideration, even when there was a conventional fixed-wing alternative. For the Royal Navy, the ability to destroy an enemy's shadowing aircraft justified the acquisition of Sea Harriers. Reconnaissance and strike roles supported the case for V/STOL, but denying targeting information to an enemy was probably the prime reason behind the purchase.

For other navies V/STOL aircraft had positive attraations, but for some a lack of pilots qualified on fast jets was a stumbling block and helicopter pilots are not necessarily good material for conversion to Sea Harrier flying. This has probably been one reason why only the Spanish and Indian navies have so far adopted the Harrier/Sea Harrier family of aircraft outside of the UK and USA, although Britain's conflict with Argentina in 1982 probably ruled out another likely customer.

It is relevant that the Soviet Navy began operating the V/STOL Yak-38 Forger from the first of the *Kiev* class of aircraft carrier in the mid-1970s. This aircraft has a two-nozzle vectored-thrust engine in the rear fuselage and two lift engines (small vertically-mounted engines that are used solely to produce jet-lift at low speeds) in the forward fuselage. Such an arrangement is arguably less efficient than having a single reliable engine, and almost certainly rules out the use of thrust vectoring in flight. The general feeling in the West is that the Yak-38

McDonnell Douglas and British Aerospace worked together to produce the Harrier II, its new wings incorporating LERXs (leading edge root extensions).

possesses only limited operational capability (having no search radar and low armament provisions), and is simply being used for extended service trials to provide data for future plans. Reports from India indicate that no serious effort has been made to market the aircraft.

After the first wave of enthusiasm for V/STOL as the only means whereby most navies could acquire a high-performance fixed-wing element, potential operators began to appreciate the price of V/STOL in terms of restricted aircraft performance. The Sea Harrier and Forger have a long range in comparison with any naval gun and most missiles, but are limited in comparison with a conventional aircraft such as an A-4 Skyhawk or Super Etendard launched by a steam catapult. A V/STOL aircraft requires around twice the thrust of a con-ventional attack aircraft, and the only way to install twice as much engine power is to cut back on disposable load (fuel and warload). It follows therefore that any V/STOL attack aircraft is penalized in comparison with its conventional equivalent. Also, in designing an engine with emphasis on a high thrust/weight ratio, economy has to be sacrificed. Fuel consumption is relatively high because the engine cannot be properly 'sized' for cruising, and thus has to be throttled back for most of the time. In

addition, the air intakes are badly oversize for cruising, and thus create spill drag. However, the V/STOL penalty would be far less in the case of a 'dogfight' aircraft, in which very high thrust is required for turning performance, climb and acceleration, but such an aircraft has yet to be produced.

In the case of the McDonnell Douglas/BAe Harrier II, warload–radius performance has taken a step forward due to a variety of devices used to enhance both VTO and STO performance, plus the use of a much larger wing of lightweight composite construction. The Harrier II still uses the Pegasus engine, however, so there is no possibility of supersonic performance above Mach 1.1 in high level flight. This may be no great problem for the US Marine Corps, which can escort its AV-8Bs with F/A-18A Hornets, but for a small navy the lack of higher supersonic capability is a drawback.

An important increase in warload–radius performance came about for the Sea Harrier with the invention of the 'ski-jump', simply a ramp at the end of the flight deck to boost the STO performance of the aircraft, enabling it to take off at lower speed, reduce deck take-off runs when not operating in VTO mode, and to take off with a higher weight of weapons and fuel. First proposed by Lt Cdr D. R. Taylor, RN, 'ski-jumps' are featured on all the Royal Navy's 'Harrier-carriers', Spain's new carrier *Principe de Asturias*, from which new AV-8Bs will operate, and *Vikrant* of the Indian Navy. For a vessel of *Vikrant*'s size – having been built as a traditional large carrier for fixed-

▲
Forgers on board Novorossiysk *with the doors over their lift engines raised.*

▶

Spanish Navy TAV-8S Matador.

wing aircraft operations – the use of a 'ski-jump' reportedly does not preclude the operation of its Alizé fixed-wing ASW aircraft. For Britain's *Invincible* class of vessel and the new Spanish carrier, it has meant no provision for catapults or take-offs by conventional fixed-wing aircraft. The lack of any carrier-borne airborne early warning aircraft during the Falklands conflict cost the Royal Navy dearly and led to the hurried development of the Westland Sea King AEW helicopter equipped with Searchwater maritime surveillance radar in an air pressurized swivelling container for operation from such carriers.

Operational procedures

It is fundamental to V/STOL aircraft (including helicopters) that they can take-off at higher weights using STO (and thus achieve better warload–radius performance) than if they use VTO, in which mode jet-lift is augmented by wing-lift. A Sea Harrier normally takes off using STO, and, having burned off its sortie fuel and expended its ordnance, lands vertically at reduced weight. Its standard operating technique is thus often referred to as STO-VL, rather than V/STOL. It is also worth noting that part of the reason why VL is the normal procedure is that the manufacturer's proposals for 180°

▲
Ski-jumping Sea Harrier.

▼
Three Sea Harriers return from a Falklands mission.

McDonnell Douglas/BAe AV-8B Harrier II

1 Nose cone
2 ARBS sensor aperture
3 Hughes angle rate bombing system (ARBS) receiver
4 All-weather landing system receiver ARN-128
5 Upper IFF aerial
6 ARBS signal processor
7 Nose pitch-control valve
8 Pitch trim unit
9 Lower IFF aerial
10 Pitot tube
11 Artificial feel system spring struts
12 Front pressure bulkhead
13 Yaw vane
14 Ram air intake (cockpit fresh air)
15 One-piece wrap-around windscreen
16 Instrument panel shroud
17 De-misting air duct
18 Rudder pedals
19 Cockpit floor level
20 Underfloor avionics bay
21 Air data computer
22 Inertial navigation system ASN-130
23 Low-voltage formation lighting strips
24 Control system linkages
25 Nosewheel doors
26 Port side console panel
27 Main fuel cock
28 Engine throttle lever
29 Nozzle-angle control lever
30 Cockpit section composite construction
31 Ejection seat
32 Canopy frame
33 Head-up display (HUD)
34 Cockpit canopy cover
35 Starboard air intake
36 Miniature detonating cord (MDC) canopy breaker
37 Sliding canopy rail
38 Ejection seat headrest
39 Ejection seat launch rails
40 Cockpit rear pressure bulkhead
41 Nosewheel bay
42 Boundary layer bleed air duct
43 Cockpit air-conditioning plant
44 Nosewheel hydraulic retraction jack
45 Nose undercarriage leg pivot fixing
46 Port air intake
47 Landing/taxiing lamp
48 Nosewheel forks
49 Nosewheel
50 Two-row intake blow-in doors

59 Engine auxiliary equipment compartment
60 Alternators
61 Rolls-Royce Pegasus (F402-RR-404A) turbofan engine
62 Oil tank
63 Forward fuel tank
64 Flush air intake
65 Engine bay venting air controls
66 Hydraulic ground servicing bay
67 Crossdam fence (forward retracting)
68 Ventral cushion augmentation strakes, port and starboard (fitted in place of gun packs)
69 Zero scarf forward nozzle fairing
70 Rotary nozzle bearing
71 Port leading-edge root extension (LERX)
72 Reaction control system bleed air ducting
73 Auxiliary power unit/gas turbine starter
74 APU intake
75 APU exhaust duct
76 Centre wing spar
77 Centre wing integral fuel tank
78 Starboard wing reaction control air duct
79 Inboard pylon
80 Fixed leading-edge fences
81 Supercritical wing section
82 Drop tank
83 Intermediate wing pylon
84 Starboard wing integral fuel tank, total internal system capacity 7500 lb (3402 kg)
85 AIM-9L Sidewinder air-to-air missile
86 Missile launch rail
87 Outboard pylon
88 Radar warning antenna
89 Starboard navigation light
90 Roll control air valves
91 Starboard formation light

102 Flap vane
103 Fuel system piping
104 Centre wing fairing panels
105 Centre wing construction
106 Water/methanol tank
107 Engine rear divider duct
108 Rear nozzle bearing
109 Centre fuselage fuel tank
110 Inboard leading-edge fence
111 Hydraulic reservoir
112 Nozzle fairing
113 Twin mainwheels
114 Rear (hot stream) exhaust nozzle
115 Titanium fuselage heat shield
116 Flap hydraulic jack
117 Wing spar/fuselage attachment joint
118 Fuel filler cap
119 Rear fuselage fuel tank
120 Aft fuselage avionics bay
121 Fuselage frame construction
122 Ram air intake
123 Avionics cooling air ducts
124 Avionics air-conditioning plant

135 Rudder
136 Rudder tab
137 Yaw control air valves
138 Tail boom
139 Tail pitch control air valve
140 Radar warning tailcone
141 Port tailplane
142 Tailplane composite construction
143 Fin attachment joint
144 Tailplane sealing plate

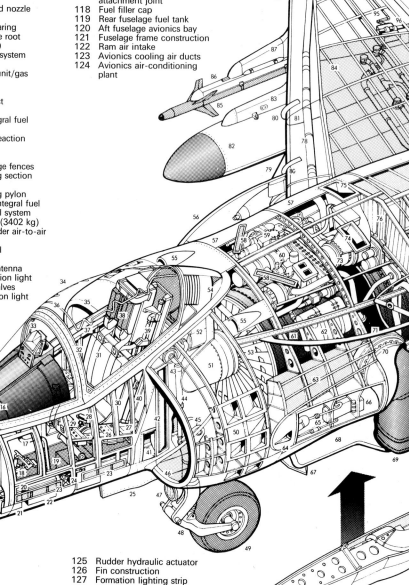

51 Intake centre fairing
52 Air-conditioning system heat exchanger
53 Engine compressor face
54 Bleed air spill ducts
55 Air-conditioning intake scoops
56 Starboard leading-edge root extension (LERX)
57 Engine equipment bay access doors
58 Upper formation lighting strips

92 Fuel jettison
93 Starboard aileron
94 Fuel jettison valve
95 Aileron/air valve interconnection
96 Aileron hydraulic jack
97 Outrigger wheel fairing
98 Starboard outrigger wheel, retracted position
99 Torque scissor links
100 Hydraulic retraction jack
101 Starboard positive-circulation, slotted flap

125 Rudder hydraulic actuator
126 Fin construction
127 Formation lighting strip
128 Fin leading-edge ribs
129 Starboard tailplane
130 Graphite/epoxy tailplane skin panels
131 Temperature probe
132 Fin-tip communications aerial
133 Radar beacon antenna
134 Rudder honeycomb construction

145 Tail bumper
146 Ventral fin
147 Lower communications aerial
148 Tailplane hydraulic jack
149 Rear fuselage main frame
150 Reaction control system air ducting
151 Fuselage frame and stringer construction
152 Lower formation lighting strip
153 Avionics bay access door
154 Airbrake hydraulic jack
155 Ventral airbrake
156 Port slotted-flap composite construction
157 Honeycomb flap vane construction
158 Outrigger wheel fairing

159 Port aileron
160 Aileron composite construction
161 Outer wing panel construction
162 Fuel jettison
163 Port formation light
164 Roll control air valve
165 Radar warning signal processor ACR-67
166 Port navigation light
167 Radar warning antenna
168 AIM-9L Sidewinder air-to-air missile
169 Missile launch rail
170 Outboard pylon
171 Drop tank, capacity 300 US gal. (1136 litres)
172 Port intermediate pylon

173 Port outrigger wheel
174 Aileron hydraulic actuator
175 Outrigger wheel leg strut
176 Hydraulic retraction jack
177 Leading-edge bleed air ducting
178 Graphite/epoxy composite wing construction
179 Outboard leading-edge fence
180 Pylon fixing hardpoint
181 Port inner pylon
182 Triple ejector bomb rack
183 Mk 82 high-explosive bombs
184 Bomb fuse controls
185 Starboard ventral gun pack
186 Port ventral gun pack
187 Ammunition tank, 100 rounds
188 Ammunition feed chute
189 Cartridge case collector box
190 30 mm cannon
191 Gun pack cushion augmentation strake
192 Gun attachment link
193 Muzzle fairing
194 Gun gas vents
195 Crossdam fence
196 Crossdam fence hydraulic jack

thrust vectoring (giving thrust reversal and thus short landing at high weights) have never been funded, in view of the cost and risks involved.

Achieving V/STOL involves squeezing the last ounce of thrust out of the engine, which is done by using a special short-term rating (15 seconds per flight). The Pegasus engine is also provided with water injection, which gives an extra 1000 lb (454 kg) thrust, taking the 'short lift, wet' rating to 21,500 lb (9750 kg). However, for many years both the RAF and the Royal Navy preferred to view water injection in STO as an emergency measure that could be invoked if the aircraft appeared to be about to strike the trees at the end of the airfield, or hit the water. In contrast, the USMC has never indicated any serious doubts regarding the use of the water injection system, and has treated the wet rating as normal. What happened to change British views of water injection was that the Indian Navy began operating Sea Harriers at extremely high ambient temperatures, requiring water injection as a standard procedure. As a result of this experience, the Royal Navy changed its attitude and began using water injection, with the philosophy that if the take-off did not proceed as planned, the aircraft's drop tanks would be jettisoned. The RAF is also now cleared to use water injection in STO mode.

The 'ski-jump' take-off may be regarded as a straightforward extension of a technique used in the early days of aviation, when taking off at high weight from a grass airfield. Rather than have his flying machine gradually disintegrate in an indefinitely long ground run, the pioneer aviator sometimes steered towards the largest bump on the field, which would jolt the aircraft off the ground, eliminating the drag on the wheels and giving it a few precious seconds in which to attain flying speed.

The ski-jump similarly gives the Sea Harrier a vertical velocity at 'unstick', allowing time for it to accelerate to a speed at which its weight can be supported by the combined effect of wing-lift and jet-lift. Since the aircraft can be 'launched' at a much lower speed than normal, take-off performance is substantially improved. For a ski-jump angle of 12 degrees, the aircraft requires roughly half the flat-deck take-off run, or can take off with an extra 3000 lb (1360 kg) of useful load.

Ski-jump trials began at the Royal Aircraft Establishment, Bedford, in August 1977, using an adjustable ramp covering a range of angles from 6° to 20°. However, when it came to modifying the Invincible class of anti-submarine carrier, it was found that there were practical restrictions on ski-jump angle. On the first two vessels of the class (Invincible and Illustrious) the position of the Sea Dart missile launcher on the forecastle restricts the ramp to 7°. In the case of the third and last ship, Ark Royal, more extensive modifications were possible, and a ski-jump angle of 12 degrees has been incorporated, giving substantially better STO performance.

Aside from the practical difficulties of incorporating a ski-jump ramp on the forward flight deck of a ship that had already been designed, the main problem in ski-jump development was to ensure that the aircraft's under-carriage was not overloaded by the curvature of the ramp. The ski-jump profile was therefore adjusted to give a gradual compression of the oleos (the equivalent of shock-absorbers on the undercarriage legs), and no

problems have been encountered in service due to ramp-induced loads.

In the case of the Sea Harrier, the pilot checks engine acceleration between 27% and 55% fan r.p.m. with the jets horizontal and the aircraft held on the brakes. The engine nozzles are lowered to the computed angle, and the flight deck officer checks this setting and that of the tailplane. The pilot confirms that the reaction control system (that will provide attitude control at low airspeeds) is being supplied with engine bleed air. He then takes off with the engine nozzles lowered 8° to keep the jets off the tailplane, and lowers them to the preset stop as the forward deck edge disappears from his field of view. Immediately after unstick the rate of climb decreases while the aircraft is accelerating forwards, but as wing-lift increases the climb rate begins to rise again and the pilot then gradually reduces the depression angle of the engine nozzles, transition to fully wingborne flight being completed at around 190 mph (305 km/h).

To land back on the ship, a decelerating transition is performed by lowering the engine nozzles and increasing thrust, bringing the aircraft to a hover alongside the selected landing spot at a height of approximately 100 ft (30 m) above the sea. The aircraft is then 'translated' to a position over the landing spot, and thrust is slightly reduced to lower the aircraft to the deck. In bad visibility the approach may be made either under control from the carrier or using data from the aircraft's radar.

V/STOL at war
The value of the Harrier V/STOL family in naval operations was a matter for considerable debate in its early service years, but these aircraft were to be subjected to genuine operational testing in the course of the conflict with Argentina over the Falkland Islands in the South Atlantic in the period April–June 1982.

On 2 April Argentine forces landed on the virtually undefended islands and took control. After a period of frantic diplomatic activity, Britain embarked a task force centred on two aircraft carriers, equipped with a total of 20 Sea Harriers and 45 helicopters. However, it was clear that the Sea Harriers would have to face 120–200 Argentine combat aircraft, including Mirages and Daggers (Israeli-built Mirage 5 types) capable of Mach 2 in level flight. Some of these might operate from the main airfield at Port Stanley (the Islands' capital), with the rest flying missions from bases in Argentina less than 500 nautical miles (925 km) away.

All possible measures to supplement the aircraft of the task force were therefore taken. As the ships approached the Falklands, eight further Sea Harriers together with nine RAF Harrier GR3s (modified to substitute for attrition Sea Harriers in the air defence role) were ferried to Ascension Island, to continue their journey south by ship. Eight Sea Harriers and six GR3s from No. 1 Sqn were taken by container ship to join the task force. It says a great deal for the flexibility of these V/STOL aircraft that they could be flown from a helicopter pad on the container ship to the carriers, and that the RAF pilots required only a day of carrier familiarization before flying their first operational missions. After reinforcements had reached Ascension, two pairs of GR3s were later flown direct to the task force and four more were taken south

A Sea Harrier begins its ski-jump take off from HMS Hermes *on dawn patrol around the Falklands, with RAF Harrier GR.3s readied on deck.*

from Ascension by another container ship, but these arrived too late to see action. Thus, a total of 28 Sea Harriers and 10 Harrier GR3s took part in operational missions, including four GR3s which had been flown direct from the UK to Ascension using in-flight refuelling, then going on to HMS *Hermes,* the 8000 mile (12,870 km) journey taking only about 18 flying hours.

The primary role of the Sea Harrier was to defend the two carriers, *Hermes* and *Invincible,* against air attack and then to provide air cover for the amphibious force that was to land on the Falklands. However, it was possible to hold the carrier battle group in comparative security 200 nautical miles (370 km) to the east of the islands for much of the conflict, hence the Harriers could be used to attack the airfield at Port Stanley, the smaller airstrips from which Pucara close-support aircraft were operating, and any surface vessels that were sent into the Total Exclusion Zone. Although initial estimates had indicated that the Royal Navy might lose one Sea Harrier per day, actual losses were far, far less than this and it was possible to switch the RAF Harriers back to ground attack operations, including close support of British ground forces following the landings on 21 May.

The initial air war took the form of a surge of activity on 1 May, followed by three weeks of comparative inactivity due to bad weather and the Argentine decision to conserve their air resources until the British landings took place. The first of three Vulcan B.2 bomber sorties carrying high-explosive bombs had taken place in the early hours of 1 May, the aim being to destroy the runway at Port Stanley airfield (to deny its use to combat and supply aircraft), and this was followed by a dawn strike by Sea Harriers, others being used for top cover.

This first major strike mission had been carried out without loss, but some attrition was anticipated when Sea Harriers came up against the Mirage types. In the event, not one Harrier of any type was lost in air combat throughout the conflict. On that first day of the air war, a total of ten Argentine Mirages and 12 Daggers were sent to sweep the skies clear for 28 A-4s and six Canberras to attack units of the task force. At first sight the Argentine aircraft had all the advantages: Mach 2 capability, Matra 530 beyond-visual-range missiles, and an excellent Westinghouse radar at Port Stanley. Nonetheless, at the end of the first day Sea Harriers had destroyed two Mirages, one Dagger and one Canberra, all without loss.

On the British side, the day's air activities confirmed numerous peacetime exercises, in which Harrier types had achieved kill ratios of 3:1, even against the latest US fighters. From that point, Argentine aircraft were to be used solely in hit-and-run missions against British ships and troops on the ground. If intercepted after the long and tiring cross-water outward flight, they would generally jettison their bombs and turn for home. The legend of 'La Muerta Negra' (The Black Death) had been born.

The Sea Harriers were operated as two squadrons, No. 800 in *Hermes* and No. 801 in *Invincible*, although the eight reinforcing aircraft (which were split equally between the two ships) had been designated as a third unit (No. 809 Sqn) in the UK. For convenience, all the GR3s were operated from *Hermes*, the larger of the two carriers. Having a longer deck with a steeper ski-jump, *Hermes* could launch aircraft at much higher weights than *Invincible*, so the latter's Sea Harriers were generally allocated the air defence sorties, while those from *Hermes* also performed ground attack and anti-surface-vessel strikes, loaded with bombs and rockets.

However, on 4 May a Sea Harrier was lost while attacking the airstrip at Goose Green, TV evidence suggesting that a 35 mm shell may have exploded in the integral fuel tank in the wing, disintegrating the aircraft and killing the pilot. In view of the very limited number of Sea Harriers available (only four were left flying in the UK, and one unfinished on the production line) and the fact that it would take three years to make replacements, it was then decided to minimize further losses. Unless a vitally important target arose, they were not to overfly well-defended areas within the defensive envelope. In the case of Port Stanley airfield, this envelope was taken to be a hemisphere of 18,000 ft (5500 m) radius, corresponding to that of the Euromissile Roland SAM (surface-to-air missile) of which one shelter version was used by Argentina against British forces. Sea Harriers would therefore only attack that airfield by toss-bombing or medium-level radar bombing. The high-risk missions were to be left to the 'more expendable' Harriers of the RAF, whose pilots were far more experienced in low-level attacks.

Sea Harriers going on combat air patrol (CAP) had a spare centreline pylon, so it became standard practice to load this with a 1000 lb (454 kg) bomb and release it over the airfield for nuisance value while on the way to the patrol area. Initially radar bombing was not permitted if the target was not visible due to cloud, which is how two aircraft came to attack the Argentine fishing vessel *Narwal* on 9 May. The boat had been used to shadow the task force, so when it was located by the Sea Harriers it was bombed and strafed until the crew abandoned ship. On the 16th two Argentine supply ships were attacked in Falkland Sound, again successfully. Sea Harriers were also involved in reconnaissance missions in preparation for the amphibious landings on the 21st.

On 18 May the eight reinforcing Sea Harriers and six GR3s were transferred to the carriers from the *Atlantic Conveyor*, boosting the fixed-wing total to 31. Aside from the loss on 4 May, two Sea Harriers had collided in bad visibility a few days later. On the 19th the eight GR3 pilots each flew two carrier familiarization sorties, and on the following day the RAF's No. 1(F) Squadron carried out its first operational mission of the conflict, destroying a fuel storage depot.

The first phase of the air war had proved not only the usefulness of the Harrier/Sea Harrier V/STOL family in both the air defence and surface attack roles, but also that the aircraft could be kept flying in appalling conditions. On some days the waves were breaking over the bow of the ship, and the aircraft on deck were covered in salt spray. At other times the humidity was close to 100% and condensation was running off the aircraft. Cockpits were inevitably wet, and moisture posed a threat to all electrical equipment, instruments and the aircraft's avionics, yet availability was maintained at a remarkable 95% by a range of ingenious protection measures.

Following the landings on 21 May, the Sea Harriers were heavily engaged in air defence missions over the amphibious landing area, while GR3s were committed to close-support duties. On that first day 800 Sqn gained victories over five A-4s and one Dagger, while 801 Sqn accounted for three Daggers and one Pucara. The aim at this stage was to form an outer defence layer beyond the reach of naval guns and guided missiles, although this put the Sea Harriers at the limit of their range, reducing time on station to 5–10 minutes. Coupled with the lack of airborne early warning aircraft, this short CAP time resulted in only a fraction of incoming Argentine aircraft being intercepted. On the other hand, once a Sea Harrier had reached a firing position behind an A-4 or Dagger, air victory seemed virtually certain, due largely to the excellent accuracy and very effective warhead of the AIM-9L Sidewinder (two were carried by each Sea Harrier).

The RAF Harrier GR3s were meanwhile performing attack missions, including the destruction of helicopters on the ground. On the 21st the squadron sustained its first loss, one GR3 being shot down by 20 mm fire and the pilot taken prisoner. Having been sent to the Falklands to support the Sea Harriers in the air defence role, the GR3s lacked some of their normal provisions for close support, and a great deal of useful equipment had been lost when the *Atlantic Conveyor* was sunk by an Exocet missile on 25 May. The RAF aircraft also lacked the equipment needed to allow their inertial systems to function effectively for navigation, when operating from a highly mobile carrier deck, although special measures by the manufacturer did allow these systems to provide vital attitude information. The GR3s received a great many small arms strikes, although only one aircraft was lost to such fire. On one occasion the reaction control ducting was holed, which started a small fire while the aircraft was landing, but damage was slight. The loss to small arms occurred on 30 May, when the fuel system of a GR3 was damaged and the pilot was obliged to ditch in the sea. As in the case of a GR3 lost to AAA on the 27th, the pilot was recovered safely. The final loss to enemy fire occurred on 2 June, when a Sea Harrier was shot down by a Roland SAM while loitering over Port Stanley.

Although most of the aluminium planking required for a Harrier runway had been lost with the *Atlantic Conveyor*, some more had been taken south in the *Stromness*, and on 5 June an 850 ft (260 m) strip was opened near the site of the first British landings. It was not possible to establish a fully-supported strip with rearming and maintenance facilities, but the ability to refuel aircraft and to hold them at readiness brought a major improvement in their effectiveness. It allowed GR3s to achieve quick reaction in close support for the first time in the conflict, it allowed Sea Harriers to stay on station for up to 30 minutes (rather than 5–10 minutes), and it permitted any aircraft that was running short of fuel to be recovered safely.

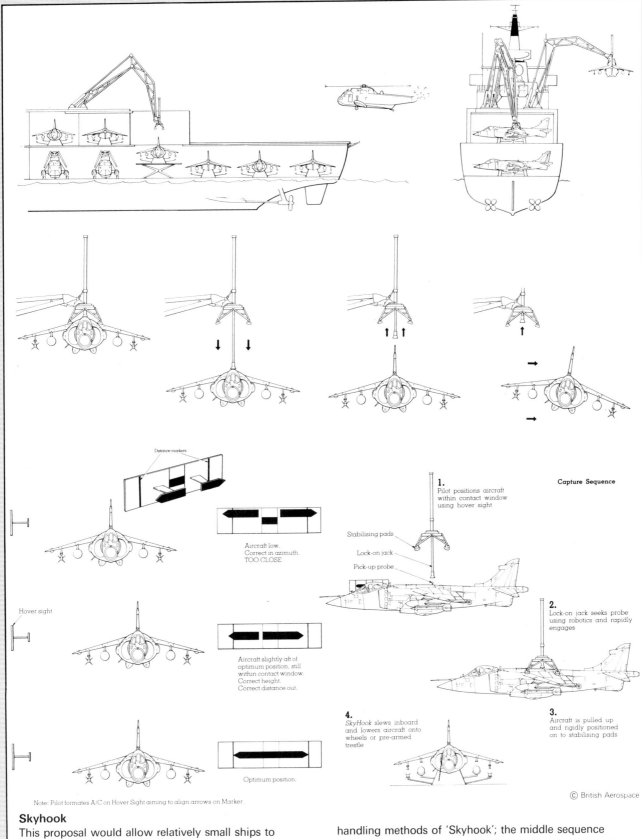

Distance markers

Aircraft low.
Correct in azimuth.
TOO CLOSE

Hover sight

Aircraft slightly aft of
optimum position, still
within contact window.
Correct height.
Correct distance out.

Optimum position.

Note: Pilot formates A/C on Hover Sight aiming to align arrows on Marker.

Capture Sequence

1.
Pilot positions aircraft
within contact window
using hover sight

Stabilising pads

Lock-on jack

Pick-up probe

2.
Lock-on jack seeks probe
using robotics and rapidly
engages

4.
SkyHook slews inboard
and lowers aircraft onto
wheels or pre-armed
trestle

3.
Aircraft is pulled up
and rigidly positioned
on to stabilising pads

© British Aerospace

Skyhook

This proposal would allow relatively small ships to operate Harrier aircraft, even in high sea states, by combining the Harrier's hovering precision with an automatically stabilized ship-mounted retrieval system. The top sequence of diagrams depicts the stowage and handling methods of 'Skyhook'; the middle sequence depicts the launch stage, once the engine is running; the bottom sequence shows how the aircraft would be retrieved, the pilot using the hover sight to ensure accurate positioning of the aircraft.

The Sea Harrier's Ferranti multi-mode coherent pulse Doppler radar for all-weather air-to-air and air-to-surface modes, capable of look-up and look-down detection of airborne targets over sea and land.

Despite these improvements, the lack of AEW aircraft continued to be a serious limitation on the effectiveness of the British air defence effort, and on 8 June Argentine aircraft made a devastating attack on British landing ships putting troops ashore at Bluff Cove. However, the RAF was also making its presence felt in the attack role, with laser-guided bombs first employed on the 13th with incredible precision. However, the following day Argentine forces on the Falklands surrendered and the war was over.

In all, the Sea Harriers had flown well over 2000 operational sorties, of which almost half had been CAPs. They had shot down 23 Argentine aircraft, without a single loss in air combat. The RAF's Harrier GR3s arrived later but they flew 125 ground attack sorties with a very small number of aircraft. Two Sea Harriers were lost to ground fire, four were lost in operational accidents (two collided, one was lost in a night launch accident, and one slid off the deck), and one was lost in the course of UK trials to clear a new ferry configuration for use in the conflict. Three GR3s were lost to ground fire, and one was severely damaged while landing on the aluminium strip, apparently due to a loss of thrust.

The conflict had validated British Aerospace claims regarding the usefulness of the Sea Harrier in the air defence role, although (as was known from previous exercises) its effectiveness depended heavily on early warning of incoming aircraft. It showed up shortcomings in the aircraft's radar, the number of air-to-air missiles carried, the lack of medium-range weapons, and endurance, matters that are now being dealt with. The conflict also highlighted the aircraft's high serviceability,

and its ability to operate in sea states and bad visibility that would probably have ruled out flying by conventional naval aircraft. However, there were also occasions on which Sea Harriers were unable to reach attacking aircraft before they released their bombs, which could be interpreted as evidence of the need for a new naval V/STOL fighter generation with more acceleration and supersonic capability.

Skyhook

One of the lessons of the Falklands conflict was that it was far better to have a Sea Harrier actually available in the combat area, rather than 200 nautical miles to the rear on the aircraft carrier. This is one reason why the idea of a Harrier 'Skyhook' for small ships is now receiving so much attention.

The idea originated with a British Aerospace test pilot, Heinz Frick, who was impressed with the ease with which a Harrier could be precisely hovered. If this hovering accuracy could be combined with an automatically-stabilized ship-mounted hook, the Harrier could return to (and presumably take off from) even a small ship in high sea states. Its disposable load would be limited, but in certain conditions that might well be acceptable.

Design studies indicated that an advanced-technology crane with a space-stabilized head can be developed, by fitting inertial sensors on the head and using their outputs to drive a hydraulic system to move the crane. The second step will be to develop a system giving visual cues, enabling the pilot to hover accurately relative to the stabilized hook. The current proposal is to have a set of three marked boards attached to the crane head and abreast of the cockpit. When the aircraft is correctly positioned, the central panel will be seen as bridging the gap between the other two, and the three parts of a painted arrow will be in line. Once the aircraft is hovering in roughly the right position, an automatic system in the crane head will complete engagement, judging the precise position of the aircraft from a pattern of IR-absorbing patches on its upper surface. The use of an IR robotic system will permit night operation. Once the weight of the aircraft is taken by the 'Skyhook', its engine can be shut down and it can be swung round and placed on the deck.

As the system is not too complicated, it follows that the 'Skyhook' concept can be made to work if there is sufficient interest among potential V/STOL aircraft operators. However, since VTOL operation places such a restriction on the effectiveness of the aircraft, it seems likely that the development of 'Skyhook' will be linked to a second-generation aircraft such as the AV-8B, rather than a Sea Harrier.

Supersonics

Although the AV-8B has brought a significant improvement in warload–radius performance over the first Harrier generation, this latest member of the family is still normally subsonic. There are no insuperable problems precluding the development of a supersonic V/STOL aircraft, but the cost would be relatively high, and there is no short-term market in sight. The Royal Navy would like to see a replacement for the Sea Harrier entering service before the end of the century, but the timescale for such an aircraft will almost certainly be dictated by the US Navy with its immense purchasing power, which is unlikely to require an F-14/FA-18 replacement before 2005/2010.

Work leading up to a supersonic V/STOL fighter is therefore proceeding slowly, with all possible power-plant concepts being examined. As far as can be established, there are currently four leading contenders in this field. Of these, the one most likely to succeed is still the single vectored-thrust engine, basically similar to the Pegasus of the Harrier family but with combustion in the front nozzles ('plenum chamber burning') to give the highest possible static thrust/weight ratio and a large thrust at supersonic speeds. In some studies the front nozzles are set low on the fuselage to prevent them from pressing the rear jets against the structure, and in others a single rear nozzle is used to reduce powerplant weight.

To minimize costs, there are strong arguments in favour of using off-the-shelf engines. Thrust matching in cruise is assisted if a relatively small engine can be given a massive thrust boost for V/STOL. Aircraft balance is easier if the mass of the powerplant is behind the aircraft CG (centre of gravity). One way which is thought to achieve all three objectives is to place a normal engine in the rear fuselage and duct the bypass air to an ejector further forward. The ejector is designed to induce additional air (admitted by large doors in the upper surface) to flow downwards, giving an augmented jet-lift. The hot gases from the engine can also be directed downward by a vectoring nozzle. In forward flight the fan flow is exhausted aft as per normal. However, although used several times in experimental V/STOL aircraft of US origin, the augmented jet-lift system has never lived up to its theoretical promise.

Alternatively, some or all of the bypass air can be ducted forwards to a vectorable nozzle, with a combustion system similar to plenum chamber burning. This concept is referred to as the remote augmented lift system (RALS).

The fourth front-runner is termed the hybrid fan. In this concept the engine has two fans separated by a long shaft, and the front fan air is ducted at low speeds to two vectorable nozzles, while auxiliary inlets open to supply air to the rear fan and main engine, the thrust of which is also vectored. This concept is attractive in principle, since it deals efficiently with a very large mass flow in the V/STOL mode, but the long shaft between the two fan stages could be a source of problems.

The third generation of V/STOL aircraft is thus many years in the future, and its powerplant could take one of several new forms.

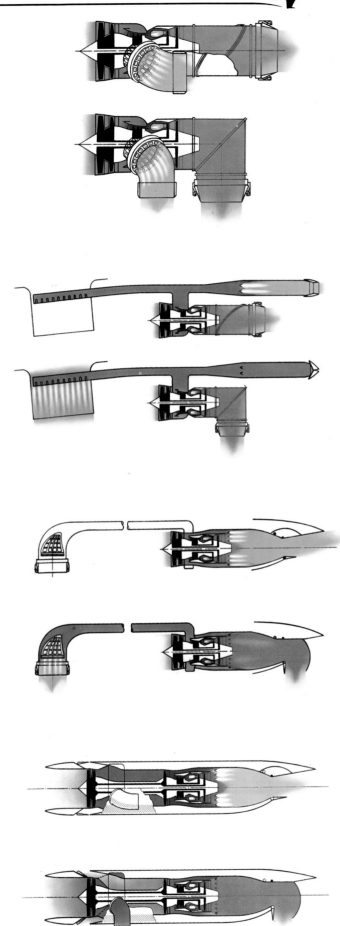

Vectored thrust with PCB (plenum chamber burning).

Ejector lift.

Remote Augmented Lift System (RALS).

Hybrid fan vectored-thrust engine.

Chapter 7
Naval air power fleets

Organic air at sea is currently dominated by the United States Navy, whose fleet of large aircraft carriers far surpasses the sea-going air power of any other nation, both in defensive and offensive capability. The US Marine Corps is well integrated with the US Navy, although its aircraft flown from land and ship are directed towards its traditional role of amphibious assault. Land-based naval operations are, however, most importantly the domain of Soviet Naval Aviation, which undertakes both tactical and strategic roles and includes in its inventory variants of all Soviet subsonic and supersonic bombers.

Having detailed the major aspects of naval air power operations and tactics in previous chapters, it is appropriate to give brief details of the world's aircraft carriers, the main foundation of organic air power at sea. However, many more naval vessels carry helicopters as standard than are covered here, including battle cruisers, cruisers, destroyers, frigates, some amphibious landing ships, depot and research vessels, and ice-breakers.

Preceding pages

HMS Illustrious recovers a Sea Harrier on completion of a successful combat air patrol.

Many types of warship other than aircraft carriers have helicopter capability. Here the French destroyer de Grasse takes on board the SA 365F Dauphin 2 test helicopter, though its normal aircraft complement is two Lynx.

The world's aircraft carriers

Argentina's ex-Colossus carrier, Veinticinco de Mayo.

ARGENTINA
Veinticinco de Mayo

First commissioned as a *Colossus* class vessel by the Royal Navy in 1945 but sold later to the Royal Netherlands Navy, it was taken over by Argentina in 1969. It has a normal aircraft complement of 22, comprising 12 Super Etendards and A-4Q Skyhawks for attack, six S-2E Trackers for ASW, and four S-61 D-4 Sea King and Alouette III helicopters for ASW and SAR/observation respectively.

Length: over 693 ft (211 m)
Power source: Conventional, offering 40,000 s.h.p.
Full-load displacement: 19,890 tons
Speed: over 24 knots
On-board weaponry: Guns.

BRAZIL
Minas Gerais

First commissioned as a *Colossus* class vessel by the Royal Navy in 1945, it was taken over by Brazil in 1956. Intended primarily for ASW but reportedly to take on an attack role at a future date, it has a normal complement of 20 aircraft, comprising perhaps eight S-2E Trackers for ASW and 12 SH-3D Sea King ASW helicopters. It is equipped with one steam catapult.

Length: 695 ft (212 m)
Power source: Conventional, offering 40,000 s.h.p.
Full-load displacement: 19,890 tons
Speed: over 25 knots
On-board weaponry: Guns

▲
Brazil's ex-Colossus class Minas Gerais.

▼
The French Navy's most important carrier, Foch.

FRANCE

Clemenceau and Foch

Commissioned in 1961 and 1963 respectively, both have the capability for atack roles but currently *Foch* is the only vessel assigned to this task, being equipped with 20 Super Etendards for attack, seven F-8E(FN) Crusader fighters, six Alizé ASW aircraft and perhaps four Etendard IVP reconnaissance aircraft, plus four Super Frelon and Alouette III helicopters. *Clemenceau* currently operates in the ASW role with about 40 helicopters. A new nuclear-powered carrier, the *Charles de Gaulle*, is scheduled to replace *Clemenceau* in 1995. Each carrier has two steam catapults.

Length: over 869 ft (265 m)
Power source: Conventional, offering 126,000 s.h.p.
Full-load displacement: 32,780 tons
Speed: 32 knots
On-board weaponry: Guns. *Foch* is to receive Naval Crotale surface-to-air missiles in place of some guns, a system already used on-board French corvettes and frigates.

Jeanne d'Arc

Commissioned for operational service in 1964, this helicopter carrier is a training vessel but could be used in wartime as a commando carrier. The current complement of four Lynx helicopters can be doubled.

Length: over 597 ft (182 m)
Power source: Conventional, offering 40,000 s.h.p.
Full-load displacement: 12,365 tons
Speed: over 26 knots
On-board weaponry: Guns and six Exocet anti-shipping missile launchers

INDIA

Vikrant

Laid down as a Royal Navy *Majestic* class carrier, HMS *Hercules* was launched in 1945 but was not commissioned until 1961, having been completed only after purchase by India in 1957 and renamed *Vikrant*. Until recently its main attack aircraft were British Armstrong Whitworth Sea Hawks, straight-winged fighters of 1950s' design and ordered by India for use on *Vikrant*. Sea Hawks are being replaced by Sea Harriers, the first batch of six having gone to No. 300 (White Tiger) Squadron in 1983. The aircraft capacity of *Vikrant* is 22, comprising Sea Harriers, Alizés for ASW, plus the Sea King helicopter for ASW. As with the Royal Navy 'Harrier carriers', *Vikrant* has now been fitted with a 'ski jump'.

Length: 700 ft (213 m)
Power source: Conventional, offering 40,000 s.h.p.
Full-load displacement: 19,500 tons
Speed: over 24 knots
On-board weaponry: Guns

ITALY

Giuseppe Garibaldi

This new light aircraft carrier was commissioned into the Italian Navy in 1985. It has been designed and constructed to permit the operation of V/STOL fixed-wing aircraft and, indeed, has a 'ski jump'. However, the Aviazione per la Marina Militare is currently an all-helicopter force. The carrier has a complement of 16 ASW helicopters, notably Agusta-Sikorsky SH-3D Sea Kings.

Length: over 591 ft (180 m)
Power source: Conventional, offering 80,000 s.h.p.
Full-load displacement: 13,320 tons
Speed: 30 knots
On-board weaponry: Four launchers for Otomat Mk 2 anti-shipping missiles, two Albatross surface-to-air missile systems, six torpedo tubes and guns. Two SCLAR Mk 2 rocket launching systems can be used for bombardment, decoying enemy anti-shipping missiles, etc.

Vittorio Veneto

This anti-submarine helicopter carrier was based on the *Andrea Doria* class cruiser but was considerably revised and lengthened, mainly to permit the operation of nine Agusta-Bell AB 212ASW helicopters for submarine search and attack, more than doubling the cruiser's helicopter complement. It was commissioned in 1969.

Length: over 589 ft (179 m)
Power source: Conventional, offering 73,000 s.h.p.
Full-load displacement: 8850 tons
Speed: 32 knots
On-board weaponry: Four launchers for Otomat Mk 2 anti-shipping missiles, Standard ER surface-to-air missiles, six torpedo tubes, guns and SCLAR rocket launchers

SPAIN

Principe de Asturias

Commissioned in 1986, this vessel is fairly typical of a new generation of 'Harrier-carriers' and, like the Royal Navy's *Ark Royal*, has a 12° 'ski jump' and airborne early warning protection in the form of Sea King helicopters equipped with Searchwater radar. It can carry 20 aircraft, comprising up to eight AV-8B Harrier IIs, and SH-3D Sea Kings and AB 212ASW helicopters.

Length: 574 ft (175 m)
Power source: Conventional, offering 46,000 s.h.p.
Full-load displacement: 15,150 tons
Speed: 26 knots
On-board weaponry: Guns

Dédalo

Commissioned as a US Navy *Independence* class carrier in 1943, but operated by the Spanish Navy since 1967, *Dédalo* has no angled deck but operates seven AV-8S Matador V/STOL aircraft and TAV-8S two-seaters, plus 20 helicopters of SH-3D Sea King, AB 212s and AH-1G HueyCobra types.

Length: 623 ft (190 m)
Power source: Conventional, offering 100,000 s.h.p.
Full-load displacement: 16,400 tons
Speed: 24 knots
On-board weaponry: Guns

UNITED KINGDOM
HMS Hermes

The last of the Royal Navy's large conventional carriers, HMS *Hermes* was commissioned in 1959. A conversion during the early 1970s saw the removal of its steam catapults, deck arrestor gear and so on, in line with its new role of helicopter-carrying commando carrier. However, this role was expanded in 1977 and in 1980 the carrier was fitted with a 'ski jump' for Sea Harriers. HMS *Hermes* was the Task Force flagship during the Falklands campaign, when its normal load of five Sea Harriers and nine Sea King helicopters was nearly trebled (including the operation of RAF Harriers from the deck). At the time of writing HMS *Hermes* was being used for training purposes.

Length: over 744 ft (227 m)
Power source: Conventional, offering 76,000 s.h.p.
Full-load displacement: 28,700 tons
Speed: 28 knots

Invincible, Illustrious and Ark Royal

These light carriers of the *Invincible* class represent the Royal Navy's main organic air element at sea and were commissioned in 1980, 1982 and 1985 respectively. HMS *Invincible* itself took a major role in the Falklands campaign, operating a far greater complement of aircraft than the normal five Sea Harriers and nine Sea Kings. The Sea Kings on-board the class now include AEW versions carrying Searchwater maritime surveillance radar, an update resulting from the Falklands experience. Another change since HMS *Invincible* was in action has been the fitting of two Phalanx and two GAM-BO1 guns. The first two vessels of the class have 7° 'ski jumps', while HMS *Ark Royal* has a 12° jump; however, *Invincible* is to undergo a refit which will also provide it with a 12° ramp.

Length: 677 ft (206 m), except HMS *Ark Royal* at nearly 686
  ft (209 m)
Power source: Conventional, offering 94,000 s.h.p.
Full-load displacement: 19,500 tons
Speed: 28 knots
On-board weaponry: Sea Dart surface-to-air missiles and
  guns

▲

The latest Invincible-class carrier is HMS Ark Royal *with a 12° ski-jump.*

▷ ▲

Port beam view of USS Dwight D. Eisenhower, *with USS* Nimitz *and the cruiser USS* South Carolina *in the background, under way during an Indian Ocean battle group turn-over operation.*

UNITED STATES OF AMERICA
Nimitz, Dwight D. Eisenhower and Carl Vinson

These CVN multi-purpose aircraft carriers are the largest vessels of their kind in the world, although not the first nuclear-powered carriers to go into service. Commissioned in 1975, 1977 and 1982 respectively, *Nimitz* and *Dwight D. Eisenhower* are assigned to the Atlantic Fleet and *Carl Vinson* to the Pacific Fleet. Each has four steam catapults and can carry up to 95 aircraft but has a normal complement of 24 Tomcat fighters in two squadrons, 34 strike aircraft as two squadrons of F/A-18s or A-7Es (24 aircraft) and A-6Es in one squadron, 16 anti-submarine aircraft as one squadron of ten Vikings and one of SH-3 Sea King helicopters, and 12 other aircraft in equal numbers of EA-6 Prowler ECM, KA-6D tanker and E-2 Hawkeye AEW types. A fourth carrier, *Theodore Roosevelt*, is due to be commissioned in 1986.

Length: 1092 ft (333 m)
Power source: Nuclear, offering 260,000 s.h.p.
Full-load displacement: 91,480 tons, except *Nimitz* at 90,940
  tons
Speed: over 30 knots
On-board weaponry: Three Basic Point Defense Missile
  Systems using Sea Sparrow missiles for surface-to-air use
  and against incoming anti-ship missiles, and guns
  Rocket launchers for decoy use.

Enterprise

This carrier was the first to be powered by nuclear reactors and was commissioned in 1961. It is the only vessel of its class and has a similar aircraft complement to the *Nimitz* class. It, too, has four steam catapults and is assigned to the Pacific Fleet.

Length: 1088 ft (332 m)
Power source: Nuclear, offering 280,000 s.h.p.
Full-load displacement: 90,970 tons
Speed: 35 knots
On-board weaponry: Three NATO Sea Sparrow Missile
    Systems (an improved Point Defense Missile System), plus
    guns
    Rocket launchers for decoy use

Kitty Hawk, Constellation, America and John F. Kennedy

By improving upon the *Forrestal* class design but retaining conventional engine power, these four vessels were produced as attack types but have since been reclassified as multi-purpose carriers. Only *Kitty Hawk* and *Constellation* are of the same length and all four differ in displacement. The four carriers were commissioned in 1961, 1961, 1965 and 1968 respectively. Each can carry an aircraft complement similar to that already detailed under an earlier entry, to a maximum of about 85 aircraft, and has four steam catapults. *Kitty Hawk* and *Constellation* are assigned to the Pacific Fleet, the other two to the Atlantic Fleet.

Data: *Kitty Hawk*
Length: 1046 ft (319 m)
Power source: Conventional, offering 280,000 s.h.p.
Full-load displacement: 81,100 tons
Speed: over 30 knots
On-board weaponry: Three NATO Sea Sparrow Missile
    Systems, and guns

Forrestal, Saratoga, Ranger and Independence

Forrestal was begun in 1952 and commissioned in 1955, making it the first post-war aircraft carrier of new construction, the other carriers in the class being commissioned in 1956, 1957 and 1959 respectively. Each can carry about 70 aircraft, launched via four steam catapults. *Forrestal* and *Saratoga* are assigned to the Atlantic Fleet, *Ranger* to the Pacific Fleet and *Independence* is undergoing modernization.

Data: *Forrestal*
Length: 1086 ft (331 m)
Power source: Conventional, offering 260,000 s.h.p.
Full-load displacement: 79,250 tons
Speed: 33 knots
On-board weaponry: Two Basic Point Defense Missile
    Systems and guns
    Rocket launcher

Midway and Coral Sea

Named after the famous carrier battles of World War II, these are among the oldest such vessels in the US Navy but have been modernized over the ensuing years and are scheduled to remain active for many more years. *Midway* was commissioned first, in 1945, followed by *Coral Sea* in 1947. Assigned to the Pacific Fleet and Atlantic Fleet respectively, each can carry about 75 aircraft. *Coral Sea* has one more steam catapult than *Midway*, with three.

Length: 979 ft (298 m)
Power source: Conventional, offering 212,000 s.h.p.
Full-load displacement: 64,000 tons *Midway*
                        65,240 tons *Coral Sea*
Speed: over 30 knots
On-board weaponry: *Midway* has launchers for Sea Sparrow,
    and both carry guns

Lexington, Bon Homme Richard and Oriskany

These three carriers were commissioned in 1943, 1944 and 1950 respectively, the disparity for the latter resulting from a temporary halt during construction following the end of the Second World War. Only *Lexington* currently remains in use, classified as an AVT (Auxiliary Aircraft Landing Training Ship) without its own aircraft. The other two are in reserve but could carry up to 80 aircraft each and have two steam catapults and gun armament (see *Hornet* and *Bennington*).

Data: *Lexington*
Length: 889 ft (271 m)
Power source: Conventional, offering 150,000 s.h.p.
Full-load displacement: 42,100 tons
Speed: over 30 knots
On-board weaponry: Guns removed

Hornet and Bennington

Commissioned in 1943 and 1944 respectively, these aircraft carriers, along with *Lexington*, *Bon Homme Richard* and *Oriskany* detailed above, were originally built as part of the *Essex* class of the Second World War. Both were put into reserve for the Pacific Fleet in 1970 (and remain so) and differ from the reserve carriers mentioned above in having two less-efficient hydraulic catapults each. The aircraft complement of each could be 45, including many helicopters.

Length: 899 ft (274 m)
Power source: Conventional, offering 150,000 s.h.p.
Full-load displacement: 40,600 tons
Speed: over 30 knots
On-board weaponry: Guns

UNION OF SOVIET SOCIALIST REPUBLICS
Kremlin

The first of up to eight huge nuclear-powered multi-purpose aircraft carriers is currently under construction on the Black Sea, having been laid down in 1983. Sea trials could begin as early as 1988, with operational deployment in the 1990s. Soviet pilots have already begun mock deck landing training on land (see Chapter 2), using MiG-29 Fulcrum and Su-27 Flanker fighters of the latest generation and other aircraft, although many believe Flanker to be the most likely of these to be found on-board ship, with possibly new versions of other aircraft already in Soviet use or of entirely new design and at present unknown in the West. It will carry specialized anti-submarine, early warning and electronic countermeasures aircraft, but Forger is unlikely to be carried except in a new form, although a second-generation V/STOL fixed-wing aircraft is another possible selection if developed.

Data: Estimated
Length: 985 ft (300 m)
Power source: Nuclear, offering perhaps 200,000 s.h.p.
Full-load displacement: 75,000 tons
Speed: over 30 knots
On-board weaponry: Extensive, but unknown

USAF F-15 Eagle and RF-4C Phantom jets fly over Minsk *during the PACAF exercise Ostfriesland II in 1983.*

Kiev, Minsk, Novorossiysk and Baku

These aircraft carriers of the *Kiev* class are basically for ASW, although their powerful on-board weaponry allows multi-purpose use including attack. They can be considered a half-way stage between the helicopter carriers of the *Moskva* class and the new nuclear-powered carriers of *Kremlin* type, although they have their own important function within the Soviet Navy. Commissioned in 1975, 1978, 1982 and 1985 respectively, each carry 12 Yak-38 Forger-A single-seat V/STOL combat aircraft, a single Forger-B two-seater, and 16 Ka-25 Hormone-A ASW helicopters and three Hormone-B target acquisition and missile guidance helicopters, or the latest Ka-27 Helix helicopters in similar numbers and roles.

Length: 896 ft (273 m)
Power source: Conventional, offering perhaps 200,000 s.h.p.
Full-load displacement: 37,100 tons
Speed: 32 knots
On-board weaponry: Four twin SS-N-12 Sandbox surface-to-surface anti-ship missile launchers with reload capability, four twin SA-N-2 Guideline /SA-N-4 surface-to-air missile launchers in *Kiev* and *Minsk* and two twin SA-N-2 and two new surface-to-air systems in the other two vessels, one twin launcher for rockets carrying nuclear depth charges, two 12-round RBU-6000 launchers, and 10 torpedo tubes.

Keeping company in the South Aegean Sea are the helicopter cruiser Moskva and the Kara-class cruiser Nikolayev.

Moskva and Leningrad

Commissioned in 1967 and 1968 respectively, these helicopter cruisers gave the Soviet Navy its first major organic air elements at sea, offering advantages in the ASW role and an increase in surface combat capability. Each carries 18 Ka-25 Hormone-A ASW helicopters, while some Hormone-As may also have air-to-surface missile capability.

Length: 620 ft (189 m)
Power source: Conventional, offering 100,000 s.h.p.
Full-load displacement: 17,500 tons
Speed: 31 kots
On-board weaponry: Two twin SA-N-3 Goblet surface-to-air missile launchers, one twin launcher for anti-submarine missiles, two 12-round RBU 6000 launchers, and guns.

Chapter 8
The world's naval aircraft

In this chapter, descriptions are given of the individual naval aircraft themselves, both the mighty that stalk the skies for aerial opposition, recce the seas for enemy warships and listen for movement under the waves, and the less-glamorous aircraft without which naval air forces would be rendered virtually impotent. Yet, despite the large number of aircraft described, only the most important trainers, transports and liaison/communications types used in a naval context can be included, such is the staggering diversity of naval aircraft.

Any grouping of this kind, however, has its anomalies. Some nations undertaking maritime air operations have no separate naval air forces but, instead, rely on their conventional land-based air forces, while other countries that have separate naval air forces may still require their conventional air forces to fly various maritime operations.

A good example of this latter case is the Royal Air Force, whose Nimrods fly a maritime patrol role. Therefore, aircraft flown by traditional land-based air forces, but which were designed for, or are used in, specific maritime roles, are included here. It should also be remembered that modern missiles give most high-performance aircraft some maritime attack capability, typified perhaps by the RAF's Hawk trainer, which was shown at the 1983 Paris Air Show with – and has demonstrated its capability of carrying – a Sea Eagle sea-skimming anti-shipping missile. Further, in-flight refuelling capability for naval aircraft is often partly and completely dependent on the tanker aircraft of conventional land air forces, although both the US Navy and Soviet Navy have their own tankers.

Aérospatiale Alouette III (France)
The Alouette III first appeared in 1959 as the Sud Aviation SE 3160, and was designed as a larger development of the Alouette II, in naval form proving an excellent platform from which to launch air-to-surface missiles or torpedoes. MAD (magnetic anomaly detector) could be carried in an ASW role. In addition to French production, of those built abroad under licence a number of Indian-built HAL Chetaks were equipped with torpedoes for the Indian Navy. Today Alouette IIIs serve on French aircraft carriers and on-board Indian warships.

Engine: One 870 s.h.p. Turboméca Artouste IIIB turboshaft, derated to 570 s.h.p.
Fuselage length: 32 ft 11 in. (10.03 m)
Main rotor diameter: 36 ft 2 in. (11.02 m)
Gross weight: 4850 lb (2205 kg)
Maximum speed: 130 mph (210 km/h)
Range: 298 miles (480 km)
Weapons: Two AS.12 wire-guided missiles, or one or two torpedoes.

Aérospatiale SA 330 Puma and AS 332 Super Puma (France)
The Puma was first developed to a requirement of the French Army, but was subsequently built in other military and civil forms, exported and produced under licence abroad. In 1978 the Super Puma appeared, a more powerful version offering an improved payload and with numerous refinements. Of the current versions available, the AS 332F is a specialized naval model (as used by Kuwait) suited to several roles including anti-shipping (ASV), ASW, and search and rescue. Features unique to this model include a folding tail-rotor pylon. For ASV and ASW missions, the AS 332F can use search radar, MAD, sonar and sonobuoys.

Data: Super Puma
Number in crew: Two or three
Engines: Two 1535 s.h.p. Turboméca Makila IA turboshafts
Length, with rotors: 61 ft 4 in. (18.70 m)
Main rotor diameter: 51 ft 2 in. (15.60 m)
Gross weight: 20,615 lb (9350 kg) with external slung load
Cruising speed: 174 mph (280 km/h)
Range without auxiliary fuel: 394 miles (635 km)

Weapons: Two AM39 Exocet anti-shipping missiles
Six AS 15TT lightweight all-weather anti-shipping missiles
One Exocet and three AS 15TTs
Two torpedoes

Aérospatiale AS 350 Ecureuil and Helibras HB 350B Esquilo (France/Brazil)
Best known as a five- or six-seat light civil helicopter, the Ecureuil has also been taken into the Royal Australian Navy for utility and survey work in AS 350B form, and the Brazilian Navy has received nine similar helicopters built by Helibras for utility purposes. A specialized armed military version of the Ecureuil has been developed as the AS 350L, armed with guns and rockets.

Number in crew: can accommodate up to six
Engine: One 641 s.h.p. Turboméca Arriel turboshaft
Fuselage length: 35 ft 10 in. (10.91 m)
Main rotor diameter: 35 ft 1 in. (10.69 m)
Gross weight: 4300 lb (1950 kg)
Cruising speed: 144 mph (232 km/h)
Range: 435 miles (700 km)

Aérospatiale SA 321 Super Frelon (France)
The Super Frelon is the largest helicopter ever built in France and, in naval terms, among the most important. Designed with the assistance of the US Sikorsky company, it was in production up to the early 1980s in commercial and military forms, the latter including the SA 321G anti-submarine helicopter for the French Navy for operation from the helicopter carrier *Jeanne d'Arc* (now using Lynx), the *Clemenceau* class aircraft carriers and for patrolling the Île Longue base of the nation's nuclear ballistic missile submarines. Twenty-four were delivered to the Navy, of which only eight are currently active in an ASW role, with a further six assigned as assault transports and two for SAR/utility. Minesweeping is another role of French Super Frelons. In addition, China operates 12 Super Frelons for ASW and Iraq has 11, some with Exocet missile capability. The ASW Super Frelon carries Doppler radar and dipping sonar.

Number in crew: Five
Engines: Three 1570 s.h.p. Turboméca Turmo IIIC6 turboshafts
Fuselage length: 63 ft 8 in. (19.40 m)
Main rotor diameter: 62 ft 0 in. (18.90 m)
Gross weight: 28,660 lb (13,000 kg)

Cruising speed: 154 mph (248 km/h)
Range: 509 miles (820 km)
Weapons: Four torpedoes or two Exocet missiles (anti-shipping role)

Aérospatiale SA 365F/AS 15TT Dauphin 2 (France)
This military helicopter was developed from the SA 365N and is suited to search and rescue (SAR), anti-surface vessel (ASV), anti-submarine (ASW), target acquisition for long-range missiles, coastal surveillance, ship escort and other maritime roles. It is armed with four AS 15TT missiles in the ASV role, when it carries Thomson-CSF Agrion radar which can track up to 10 targets simultaneously while searching for others. For ASW, MAD is added to the equipment, with options for sonar, torpedoes and sonobuoys. The Saudi government ordered 24 of these helicopters in 1980, 20 for ASV/ASW with MAD on board for use from frigates and shore and four for SAR, while Ireland ordered five for maritime reconnaissance, SAR and fishery surveillance with Bendix RDR L500 search radar.

Number in crew: Two
Engines: Two 700 s.h.p. Turboméca Arriel 520M
   turboshafts
Fuselage length: 39 ft 9 in. (12.11 m)
Main rotor diameter: 39 ft 2 in. (11.93 m)
Gross weight: 8818 lb (4000 kg)
Cruising speed: 156 mph (252 km/h)
Range: 558 miles (898 km)
Weapons: As detailed above

Agusta A 109A Mk II (Italy)
Among the civil and military versions of this helicopter, a naval model provides for ASV, ASW, patrol, mid-course missile guidance, reconnaissance, SAR, and electronics and other roles. Able to operate by day or night, the naval A-109A in ASV form carries search radar and can mount an attack using AS.12 or AM-10 air-to-surface missiles. For ASW the helicopter carries MAD and up to two torpedoes and six marine markers.

Number in crew: Two or three
Engines: Two 400 s.h.p. Allison 250–C20B turboshafts
Fuselage length: 35 ft 2 in. (10.71 m)
Main rotor diameter: 36 ft 1 in. (11.00 m)
Gross weight: 5732 lb (2600 kg)
Cruising speed: 178 mph (287 km/h)
Range: 368 miles (593 km)
Weapons: As detailed above

Agusta-Bell 212ASW (Italy)
The Italian company Agusta has built US Bell helicopters under licence for many years, including a version of the Bell Model 212. Agusta has also developed its own anti-submarine helicopter from the AB 212, which has the designation AB 212ASW but which is equally suited to anti-surface vessel missions and search and rescue. The main form of detection in an ASW role is dipping sonar, attacks being mounted using two torpedoes or depth charges. For ASV, long-range search radar is used, with two Sea Skua or Marte Mk 2 air-to-surface missiles performing the attack. Like the A 109A Mk II, the AB 212ASW can also perform mid-course guidance to ship-fired Otomat 2 anti-shipping missiles. AB 212ASWs are in widespread service, to be found on board many Italian Navy vessels and within the naval strengths of other nations including Greece, Iran, Iraq, Peru, Spain, Turkey and Venezuela.

Number in crew: Three or four
Engine: One 1875 s.h.p. Pratt & Whitney Canada
   PT6T-6 Turbo Twin Pac turboshaft
Fuselage length: 42 ft 5 in. (12.92 m)
Main rotor diameter: 48 ft 2 in. (14.69 m)
Gross weight: 11,176 lb (5070 kg)
Maximum speed: 122 mph (196 km/h)
Range with auxiliary fuel: 414 miles (667 km)
Weapons: As detailed above

Preceding pages
Tupolev Tu-142 Bear-D.

Royal Navy Sea King HAS.5s of No 820 Squadron being ranged to the flight deck of HMS Illustrious.

▲
Alouette III.

◀
Royal Australian Navy AS 350B Ecureuil lands on-board the frigate Adelaide.

▷▲
French Navy SA 321G Super Frelon, the largest helicopter built in France to date.

▷▼
Agusta-Sikorsky HH-3F (S-61R) with its hydraulically operated rear loading ramp lowered, used by the Italian Air Force for search and rescue.

Agusta-Sikorsky AS-61 and ASH-3D/H, Sikorsky SH-3H and Westland Sea King (Italy/USA/UK)

Agusta has built the US Sikorsky S-61/SH-3 helicopter under licence since 1967 and has provided the Italian Navy and others with examples for various roles including anti-submarine, anti-surface vessel, search and rescue, anti-surface vessel missile defence (ASMD), and VERTREP (vertical replenishment, or the hauling of supplies between shore/ship and ship). The current ASW version is the ASH-3H, which is similar to the US Navy's SH-3H built by Sikorsky (now out of production), although with more power than the SH-3H's two 1400 s.h.p. General Electric T58-GE-10 turboshafts (the earlier SH-3D ASW helicopter is in US Navy Reserve). In Britain Westland too has built the Sikorsky S-61/SH-3 under licence for many years and, like the others, has put ASW and other versions into service with its own armed forces and has exported widely. The latest Royal Navy version is the HAS Mk 5, fitted with two 1660 s.h.p. Rolls-Royce Gnome H.1400-1 turboshaft engines and used for ASW and SAR while carrying dipping sonar, AW 391 search radar in a dorsal radome, and other equipment. The Royal Navy and Spain also operate AEW variants of the Sea King, offering early warning against attack upon carrier groups (in each case centred upon Searchwater maritime surveillance radar with 360° scan). Agusta ASH-3Hs have a 'chin' radome. Utility versions of the helicopter are also in naval service, while ASW and similar models can carry troops, cargo, etc. if required.

Data: Agusta-Bell ASH-3H
Number in crew: Four
Engines: Two 1500 s.h.p. General Electric T58-GE-100
    turboshafts
Fuselage length: 54 ft 9 in. (16.69 m)
Main rotor diameter: 62 ft 0 in. (18.90 m)
Gross weight: 21,000 lb (9525 kg)
Cruising speed: 138 mph (222 km/h)
Range: 725 miles (1166 km)
Armament: Up to four torpedoes
           Four depth charges
           Four AS.12 or two Exocet or Marte Mk 2
           air-to-surface missiles (Indian Navy
           Westland Sea Kings can carry Sea Eagle)

Antonov An-12 (USSR)

Known to NATO by the reporting name Cub-B, ten examples of an ELINT (electronics intelligence) variant of the An-12 cargo transport aircraft are in service with the Soviet Naval Air Force. ECM (electronic counter-measures) Cub-Cs and Ds are operated both by the Air Force and Naval Air Force, while a new Cub is undergoing trials as an anti-submarine aircraft.

Engines: Four 4000 e.h.p. Ivchenko AI-20K turboprops
Length: 108 ft 7 in. (33.10 m)
Wing span: 124 ft 8 in. (38.00 m)
Gross weight: 134,480 lb (61,000 kg)
Maximum speed: 482 mph (776 km/h)
Range: 3540 miles (5700 km)
Weapons: Cubs carry two 23 mm guns in a tail turret

Beechcraft Maritime Patrol B200T (USA)

This maritime patrol version of the Super King Air 200 light commercial transport has many possible roles, which can include search and rescue, surface and subsurface monitoring using 360° scan search radar or sonobuoys and processor, and offshore inspection patrols. Operators include Algeria, Chile, Japan, Peru and Uruguay.

Engines: Two 850 s.h.p. Pratt & Whitney Canada
    PT6A-42 turboprops
Length: 43 ft 9 in. (13.34 m)
Wing span over removable wingtip fuel tanks: 56 ft
    7 in. (17.25 m)
Gross weight: 14,000 lb (6350 kg)
Typical patrol speed: 161 mph (259 km/h)
Range: 2061 miles (3317 km)

Bell Model 209 HueyCobra and SeaCobra (USA)

In 1965 Bell flew the prototype of an entirely new type of helicopter, a very slender tandem two-seater designed specifically for attack. It was the first of its kind in the world and was hurriedly taken into US Army service as the AH-1G HueyCobra. From 1970 the US Marine Corps began receiving a twin-engined version of the HueyCobra designated AH-1J SeaCobra, while Spain received eight HueyCobras under the Spanish military designation Z.14 for anti-shipping strikes from its aircraft carrier (four currently used). In 1977 the USMC began receiving an improved version of the SeaCobra as the AH-1T, bringing the current total AH-1J/T strength to about 72.

Data: AH-1T Improved SeaCobra.
Number in crew: Two
Engines: One 1970 s.h.p. Pratt & Whitney Canada
    T400-WV-402 coupled turboshaft
Fuselage length: 48 ft 2 in. (14.68 m)
Main rotor diameter: 48 ft 0 in. (13.41 m)
Gross weight: 14,000 lb (6350 kg)
Maximum speed: 207 mph (333 km/h)
Range: 261 miles (420 km)
Weapons: General Electric 'chin' turret with 20 mm
          M197 three-barrel gun plus:
          TOW or Hellfire anti-armour air-to-
          surface missiles.
          Four rocket pods, two M118 grenade
          launchers, or two Miniguns or other
          weapons.

Bell and Agusta-Bell Models 204 and 205, and Bell UH-1N (USA/Italy)

When Bell developed its Model 204 for the US Army in the late 1950s, which became the Iroquois, it quickly became one of the classic aircraft of aviation history. It was built in the USA and abroad, and variants included some for naval use including the roles of assault/transport (US Marine Corps in particular – some armed), helicopter training and SAR. The Italian company, Agusta, also developed the AS 204AS for anti-submarine work. From the Model 204, Bell developed the longer Model 205 general-purpose helicopter, which has also been built in Italy and Japan, and this too has found its way into naval service. The Models 204 and 205 were built as single-engined helicopters, the Model 205's 1400 s.h.p. Avco Lycoming T53-L-13 turboshaft allowing it to carry up to 14 troops or 3880 lb (1759 kg) of cargo. A twin-engined version of the UH-1 Iroquois appeared subsequently as the Bell Model 212 Twin Two-Twelve, which entered service with the US Navy and Marine Corps as the UH-1N and with the Canadian Armed Forces as the CH-135.

Data: UH-1N
Engines: One 1800 s.h.p. Pratt & Whitney Canada
    PT6T-3B Turbo Twin Pac, comprising two
    turboshafts
Fuselage length: 42 ft $4\frac{3}{4}$ in. (12.92 m)
Main rotor diameter: 48 ft $2\frac{1}{4}$ in. (14.69 m)
Gross weight: 11,200 lb (5080 kg)
Maximum speed: 115 mph (185 km/h)
Range: 261 miles (420 km)

Beriev Be-6 (USSR)
Code named Madge by NATO, the Be-6 is a maritime
reconnaissance flying-boat of Soviet design that dates
back to the 1950s. Despite its age, eight are thought to be
operated by the Aviation of the People's Navy of China. A
retractable radome is carried under the hull, and each
could be equipped with MAD (see Chapter 3). Others may
be in storage, as the Navy was thought to have 20 Be-6s in
1980.

Engines: Two 2300 s.h.p. Shvetsov ASh-73TK radials
Length: 77 ft 1 in. (23.50 m)
Wing span: 108 ft 3 in. (33.00 m)
Gross weight: 51,710 lb (23,455 kg)
Maximum speed: 258 mph (415 km/h)
Range: 2980 miles (4800 km)
Weapons: Four or five 23 mm cannon plus:
    perhaps 4400 lb (2000 kg) of ASW
    weapons

Beriev M-12 Tchaika (USSR)
See p. 69 for data.

**Boeing B-52G Stratofortress, and E-3 Sentry and Tupolev
Tu-126** (USA/USSR)
Currently the USAF's only fully operational heavy long-
range bomber, the B-52 is best known in its strategic role
while armed with air-launched cruise missiles (ALCMs)
and SRAM missiles. Those B-52Gs not assigned to be
modernized to carry ALCMs have replaced old B-52Ds in
a maritime support role, one squadron of 15 aircraft
assigned to Pacific operations and another squadron for
the Atlantic. The main weapon for this mission is the
Harpoon all-weather anti-shipping missile with a range
of more than 50 nautical miles (carrying a high-explosive
blast warhead at subsonic speed.) To assist in over-the-
horizon target location and tracking, a small number of
USAF Boeing E-3 Sentry AWACS (airborne warning and
control system) aircraft are being modified under
programme Outlaw Shark. Apart from the modified
AWACS aircraft for the role mentioned, many operational E-3s have considerable maritime surveillance
capability and, like all E-3s, carry AN/APY-1 surveillance
radar in a rotating rotodome mounted above the fuselage.
The Soviet air force equivalent of the E-3 is the Tupolev
Tu-126 (NATO Moss), again with a rotodome and
reportedly most successful when operating over water.
However, the main tasks of the E-3 and Moss are to detect,
track and identify potential enemy aircraft and missiles
and support friendly fighters during intercept or friendly
strike forces in an attack.

Data: B-52G
Number in crew: Six
Engines: Eight 13,750 lb (6237 kg) thrust Pratt &
    Whitney J57-P-43WB turbojets
Length: 160 ft 11 in. (49.05 m)
Wing span: 185 ft 0 in. (56.39 m)
Maximum speed: 595 mph (957 km/h)
Range: 7500 miles (12,070 km)
Weapons: Four 0.50-in. machine-guns in a remotely-
    operated tail turret plus:
    12 Harpoon air-to-surface anti-shipping
    missiles

*The Soviet AWACS aircraft known to NATO as Moss is
reportedly most successful when operating over water.*

◁ ▲
US Marine Corps Bell AH-1T SeaCobra carrying TOW missiles.

◁ ▼
Bell UH-1H Iroquois, flown on SAR missions by the Brazilian Air Force.

▼
French Navy Alizés with their radomes retracted.

▲
Royal Air Force Shackleton AEW.2 flies alongside the replacement airborne early warning aircraft, the Nimrod AEW.3.

▼
In a simulated mission, a Boeing Surveiller directs a Boeing hydrofoil to investigate a suspicious vessel. Equipped with side-looking airborne modular multi-mission radar produced by Motorola and capable of 'seeing' 100 nautical miles to each side of the aircraft, it normally cruises at 30,000 ft.

Boeing Vertol H-46 Sea Knight and Kawasaki KV107II/IIA (USA/Japan)
First flying in 1962, the Boeing Vertol Model 107 was put into production for the US Navy and Marine Corps, and between 1964 and 1971 no fewer than 624 were delivered. The CH-46 Sea Knight (in various models) is the Marine Corps' standard assault helicopter, accommodating up to 25 troops or 4000 lb (1814 kg) of cargo. With full load it has a combat radius of 100 nautical miles. Currently, a programme is still under way to bring 273 USMC Sea Knights up to the latest CH-46E standard, with more powerful engines and various improvements to increase survivability and crash resistance. The UH-46 Sea Knight is the US Navy's VERTREP helicopter, used in the transfer of supplies between ships but also able to undertake SAR if required. For many years Kawasaki of Japan has held manufacturing rights for the Boeing Vertol 107 Model II and has produced both commercial and military versions. The JMSDF has received KV107II/IIA-3s for mine countermeasures, the IIA model representing the more powerful version of the two and all serving with the 111th Air Wing, while the Swedish Navy uses KV107IIs as HKP 4Cs in an anti-submarine and utility role. HKP 4Cs are powered by Rolls-Royce Gnome H.1200 turboshaft engines. The Canadian Armed Forces use US-built CH-113 Labradors for SAR.

Data: CH-46E
Engines: Two 1870 s.h.p. General Electric T58-GE-16
    turboshafts
Data: CH-46D/F
Engines: Two 1400 s.h.p. General Electric T58-GE-10
    turboshafts
Fuselage length: 44 ft 10 in. (13.66 m)
Rotor diameter (each): 51 ft 0 in. (15.54 m)
Gross weight: 23,000 lb (10,433 kg)
Maximum speed: 165 mph (266 km/h)
Range with a 4550 lb (2063 kg) payload: 238 miles
    (383 km)

Breguet Alizé (France)
First flown in 1956, the Alizé is a carrier-borne anti-submarine aircraft which entered French Navy service three years later. Although many are not currently to be found on-board ship, the Navy has 34 remaining of which 28 have recently been modernized to keep them operational into the next decade. This work included replacing the old search radar carried in a retractable underfuselage radome with new Thomson-CSF Iguane, and ESM equipment. The Indian Navy also flies Alizés, four on board *Vikrant*.

Number in crew: Three
Engine: One 2100 e.h.p. Rolls-Royce Dart RDa 7 Mk 21
    turboprop
Length: 45 ft 6 in. (13.86 m)
Wing span: 51 ft 2 in. (15.60 m)
Gross weight: 18,078 lb (8200 kg)
Maximum speed: 292 mph (470 km/h)
Range: 1550 miles (2500 km)
Weapons: Weapons bay for:
        one torpedo or three 160 kg depth charges
        Underwing racks for:
        two depth charges and six 5-in. rockets or
        two AS. 12 air-to-surface missiles.

British Aerospace 748 and Boeing Surveiller (UK/USA)
A maritime patrol version of the BAe 748 twin-turboprop airliner is available for ASW and ASV as the Coastguarder, carrying Litton APS-504(V)3 search radar in an underfuselage radome, ESM, MAD and much other equipment and weapons. Two examples of the BAe 748 are used by the Royal Australian Navy carrying navigational and electronic training equipment and are flown in an ECM role. Separately, Boeing has produced a maritime surveillance version of its Model 737–200 airliner, equipped with side-looking multi-mission radar. Known as the Surveiller, three are operated by the Indonesian Air Force. They can also be used to transport 102 passengers.

Data: Surveiller
Engines: Two Pratt & Whitney JT8D turbofans
Length: 100 ft 2 in. (30.53 m)
Wing span: 93 ft 0 in. (28.35 m)

British Aerospace AV-8 Harrier, AV-8S Matador and Sea Harrier (UK)
See p. 48 for data.

British Aerospace Nimrod (UK)
Designed as a Hawker Siddeley Shackleton replacement (although six Shackleton AEW.2s continue in RAF service as airborne early warning aircraft, to be superseded soon by Nimrod AEW.3s), the Nimrod is the RAF's standard maritime patrol aircraft. Three also serve as Nimrod R.2 ELINT (electronic intelligence) aircraft. Based upon the Comet 4C airliner airframe, but with an unpressurized lower pannier carrying the operational equipment and weapons, it carries either EMI ASV-21D or Searchwater search radar, MAD in a tail 'sting', a searchlight, cameras, EWSM (electronic warfare surveillance measures) gear, sonobuoys and so on for its main roles of anti-submarine, anti-shipping, long-range reconnaissance, and day and night photography. The RAF currently has 31 maritime patrol Nimrods in use.

Number in crew: 12
Engines: Four 12,140 lb (5506 kg) thrust Rolls-Royce
    RB168-20 Spey Mk 250 turbofans
Length: 126 ft 9 in. (38.63 m)
Wing span: 114 ft 10 in. (35.00 m)
Gross weight: 177,500 lb (80,510 kg)
Maximum speed: 575 mph (926 km/h)
Ferry range: up to 5755 miles (9265 km)
Weapons: Torpedoes, bombs, mines, two Sidewinder
        air-to-air missiles for self-defence, cannon
        pod, or Harpoon air-to-surface missiles

CASA C-212 Series 200 Aviocar (Spain)
CASA has available, like many companies producing civil twin-turboprop light transports, specialized versions of the C-212 for ASW and maritime patrol purposes. The ASW model carries search radar under the fuselage with 360° scanning, ESM (electronic support measures), MAD, sonobuoys and processor, and weapons among its equipment, while the maritime patrol model has 270° scan search radar in the nose, camera, searchlight and perhaps FLIR if required. Operators are Spain, Sudan, Uruguay and Venezuela.

Number in crew: Five or more
Engines: Two 900 s.h.p. Garrett TPE331-10R-511C
    turboprops
Length: 49 ft 9 in. (15.16 m)
Wing span: 62 ft 4 in. (19.00 m)
Gross weight: 18,519 lb (8400 kg)
Cruising speed: 219 mph (353 km/h)
Range: 1900 miles (3055 km)
Weapons: ASW model can carry torpedoes, rockets
    and other stores

Canadair CL-215 (Canada)
One of the few amphibious flying-boats built today, the
remarkable CL-215 is best known as a firefighting
waterbomber but is fully suited to other roles including
transport, maritime patrol and search and rescue (SAR).
Apart from those aircraft used in a non-naval context, the
Hellenic Air Force operates CL-215s for SAR and the
Royal Thai Navy undertakes SAR and patrol missions.
For SAR, special equipment includes search radar.

Number in crew: Six
Engines: Two 2100 s.h.p. Pratt & Whitney R-2800-CA3
    radials
Length: 65 ft 0½ in. (19.82 m)
Wing span: 93 ft 10 in. (28.60 m)
Gross weight: 43,500 lb (19,731 kg)
Cruising speed: 181 mph (291 km/h)
Range: 1300 miles (2094 km) with a 3500 lb (1587 kg)
    internal load

Dassault-Breguet Atlantic (France)
See p. 69 for data.

Dassault-Breguet Etendard and Super Etendard (France)
Super Etendard transonic strike fighters have been much
in the news over recent years. Armed with Exocet
missiles and other weapons, both Argentina and Iraq
have demonstrated their capability to attack shipping
using small numbers of this aircraft. The original
Etendard was designed in the 1950s as a land-based
strike fighter for NATO and the French Air Force but, in
the event, only the French Navy took production
examples, mainly for carrier operations. Of these, the
French Navy has eight Etendard IVPs remaining for
possible use as reconnaissance aircraft from its aircraft
carriers, while 12 Etendard IVMs are based with an
operational conversion unit and represent strike con-
figured aircraft. The IVM is powered by a 9700 lb (4400
kg) thrust Atar 8B turbojet engine. When the French
Navy required an updated version, the higher-powered
Super Etendard was developed, which also has
modifications to its wings, and more sophisticated
avionics and armament. Indeed, the Super Etendard may
look similar to the Etendard but it represents almost a
new design. The first Super Etendard was taken into
service in 1978, and production ended in 1983 after the
French Navy had received 71 and 14 had gone to
Argentina. Five French aircraft were subsequently leased
to Iraq. To increase further the strike capability of the
French Navy, 50 Super Etendards are receiving the new
ASMP air-to-surface missile (with a 100–150 kt nuclear
warhead) for attacking heavily defended targets.

Data: Super Etendard
Number in crew: One
Engine: One 11,025 lb (5000 kg) thrust non-
    afterburning SNECMA Atar 8K-50 turbojet
Length: 46 ft 11½ in. (14.31 m)
Wing span: 31 ft 6 in. (9.60 m)
Gross weight: 26,455 lb (12,000 kg)
Maximum speed: about Mach 1
Combat radius: 530 miles (850 km)
Weapons: Two 30 mm DEFA guns plus:
    Two Matra Magic air-to-air missiles
    One Exocet anti-ship missile and auxiliary
        fuel tank
    One ASMP nuclear air-to-surface missile
    Four 250 kg or 400 kg bombs
    Four rocket pods

Dassault-Breguet Gardian (France)
See p. 81 for data.

Dassault-Breguet HU-25A Guardian (France)
Under the designation HU-25A, the US Coast Guard has
received 41 examples of the Mystère-Falcon 20G,
prepared as medium-range surveillance aircraft. All were
delivered in 1982–83 and are now based at nine stations.
Specialized equipment includes search and weather
radar and optional side-looking airborne radar (SLAR),
infra red/ultra violet scanners, forward-looking infra-red
(FLIR), a reconnaissance camera and a television camera
with laser illumination.

Number in crew: Five to seven
Engines: Two 5440 lb (2467 kg) thrust Garrett ATF3-6-
    2C turbofans
Length: 56 ft 3 in. (17.15 m)
Wing span: 53 ft 6 in. (16.30 m)
Gross weight: 33,510 lb (15,200 kg)
Cruising speed: 531 mph (855 km/h)
Range: 2590 miles (4170 km)
Weapons: Attachment points under the fuselage and
    wings for equipment and weapons

De Havilland Canada DHC-5D Buffalo (Canada)
This STOL transport is in service around the world, being
particularly useful where airstrips dictate aircraft with
short take-off and landing runs; with a 12,000 lb (5443 kg)
payload the Buffalo can take off in 950 ft (290 m) and land
in just 550 ft (168 m). Most Buffalos in military service are
operated by conventional land air forces but some are in
Navy hands. The maximum payload for a STOL mission
is 18,000 lb (8164 kg), which can include 34 personnel.
The Canadian Armed Forces operates Buffalos for both
transport and search and rescue missions, while the
Mexican Navy also uses the type.

Number in crew: Three
Engines: Two 3133 s.h.p. General Electric CT64-820-4
    turboprops
Length: 79 ft 0 in. (24.08 m)
Wing span: 96 ft 0 in. (29.26 m)
Gross weight: 49,200 lb (22,316 kg)
Typical cruising speed: 290 mph (467 km/h)
Range with maximum payload: 691 miles (1112 km)

▶

Royal Thai Navy Canadair CL-215 amphibian.

▼

Canadian Armed Forces, Air Transport Command, de Havilland Canada Buffalo used for both transport and SAR duties.

De Havilland Canada DHC-6-300MR (Canada)
Based upon the Twin Otter, the DHC-6-300MR is a maritime reconnaissance aircraft with undernose search radar, a wing-mounted searchlight, and three stores attachment points for various light attack weapons including guns and rockets. The main cabin can accommodate up to 20 persons or six stretchers.

Number of crew: Four or five
Length: 51 ft 9 in. (15.77 m)
Wing span: 65 ft 0 in. (19.81 m)
Gross weight: 14,000 lb (6350 kg)
Typical cruising speed: 193 mph (311 km/h)
Range with two cabin auxiliary fuel tanks: 1681 miles (2705 km)
Weapons: See above

EMBRAER EMB-111 (Brazil)
Another land-based maritime patrol aircraft developed from a small commercial transport (the EMB-110 Bandeirante), the EMB-111 was put into production for the Brazilian Air Force's Coastal Command as the P-95. The search radar is housed in an elongated nose and

wingtip fuel tanks extend range. Chilean Navy aircraft also carry Thomson-CSF passive ECM. Searchlight and flares are among options available to operators, while some P-95s carry smoke grenade markers.

Number of crew: Up to seven
Engines: Two 750 s.h.p. Pratt & Whitney Canada PT6A-34 turboprops
Length: 48 ft 11 in. (14.91 m)
Wing span: 52 ft 4 in. (15.95 m)
Gross weight: 15,432 lb (7000 kg)
Cruising speed: 223 mph (360 km/h)
Range: 1830 miles (2945 km)
Weapons: Eight 5 in. or 28 × 2.75 in. rockets or other weapons

Fokker F.27 Maritime and Maritime Enforcer (The Netherlands)
See p. 81 for data.

Government Aircraft Factories Searchmaster (Australia)
See p. 81 for data.

▲
De Havilland Canada DHC-6-
300MR maritime reconnaissance
Twin Otter, with a chin radome
for its Litton radar, searchlight
pod with 50 million candlepower,
and a gun and two rocket pods
on the remaining underwing
attachment points.

◄

Brazilian Air Force's Coastal
Command Embraer P-95, armed
with 5-in. HVAR air-to-surface
rockets.

▶
Grumman TC-4C Academe flying
classroom variant of the
Gulfstream I.

Flight crew and pilot prepare to fire the engines of an A-6E Intruder on USS Coral Sea.

Grumman A-6 Intruder and TC-4C Academe (USA)

Specially developed as a carrier-borne low-level attack bomber to carry conventional or nuclear weapons to targets by day or night and in all weathers, the Intruder has been in US Navy service since 1963. The current version is the A-6E, and these carry as part of their electronics a package standing for target recognition and attack multisensor (TRAM) and including both infra-red and laser equipment. Some 250 are in current service, equipping 18 Navy and Marine Corps squadrons plus a further three operational training squadrons. Training of Intruder bombardier/navigators is undertaken on Grumman TC-4C Academe aircraft, a special version of the Gulfstream transport.

Data: A-6E
Number of crew: Two
Engines: Two 9300 lb (4218 kg) thrust Pratt &
  Whitney J52-P-8B turbojets
Length: 54 ft 9 in. (16.69 m)
Wing span: 53 ft 0 in. (16.15 m)
Gross weight, for catapult take-off: 58,600 lb
  (26,580 kg)
Maximum speed: 644 mph (1037 km/h)
Range: 1011–3245 miles (1627–5222 km)
Weapons: Up to 18,000 lb (8165 kg) of weapons
        including:
        Harpoon air-to-surface anti-shipping
        missiles
        28 × 500 lb (227 kg) bombs
        Three 2000 lb (907 kg) bombs plus auxiliary
        fuel tanks
        Sidewinders can be carried for self-defence

Grumman C-2A Greyhound (USA)

Based upon the airframe of the Hawkeye, the Greyhound is unique in being a COD (carrier on-board delivery) transport, intended to carry up to 10,000 lb (4536 kg) of cargo or up to 32 passengers between the shore and an aircraft carrier at sea. The aircraft's cargo payload can increase by 50% for land-to-land missions. It can be catapulted from the deck and make arrested landings if necessary. Only 12 of the original Greyhounds remain in service with the US Navy, but 39 new aircraft are expected to be built.

Data: Existing C-2As
Number in crew: Three
Engines: Two 4050 e.h.p. Allison T56-A-8A turboprops
Length: 56 ft 8 in. (17.27 m)
Wing span: 80 ft 7 in. (24.56 m)
Gross weight: 54,830 lb (24,870 kg)
Maximum speed: 352 mph (567 km/h)
Range: 1650 miles (2660 km)
Weapons: None

Grumman E-2 Hawkeye (USA)

See p. 95 for data.

Grumman EA-6B Prowler (USA)

See p. 98 for data.

Grumman F-14 Tomcat (USA)

The US Navy's main carrier-based multi-role fighter, the Tomcat first became operational in 1972 in the initial F-14A version. Twin finned, the aircraft's most prominent aerodynamic feature is its variable-geometry wings, which a computer automatically positions for the best aerodynamic efficiency in flight, even during high-g manoeuvres. The AWG-9 weapons control system has track-while-scan radar which can control simultaneous attacks on six different targets while tracking up to 18 others, its large size allowing it to detect aircraft at distances of 100 nautical miles range and across a similar scan sector. This long-range detection and tracking capability is matched by the long range of the Tomcat's Phoenix missiles. The US Navy has received about 500 F-14As, but those delivered in fiscal year 1987 will include the first F-14Ds with improved avionics and weapon systems plus new General Electric F110-400 engines.

Data: Latest F-14A version
Number in crew: Two
Engines: Two 20,900 lb (9480 kg) thrust with
  afterburning Pratt & Whitney TF30-P-414A
  turbofans
Length: 62 ft 8 in. (19.10 m)
Wing span: 64 ft 1½ in. (19.54 m) spread
        38 ft 2½ in. (11.65 m) swept
Gross weight: 74,349 lb (33,724 kg)
Maximum speed: Mach 2.34
Maximum range with internal and auxiliary fuel: 2000
  miles (3220 km)
Weapons: One 20 mm M61 six-barrel gun in the nose
        plus:
        Four Sparrow and four Sidewinder air-to-
        air missiles
        Four Phoenix and four Sidewinder missiles
        Six Phoenix and two Sidewinder missiles
        Six Sparrow and two Sidewinder missiles
        Up to 14,500 lb (6577 kg) of attack weapons
        and air-to-air missiles

▷▲

A stunning view of an F-14 as it crosses the stern of USS John F. Kennedy *and lands on.*

▷

The Grumman C-2A Greyhound is the US Navy's logistical lifeline to the carrier battle groups.

Grumman HU-16 Albatross (USA)
First flown in 1947 and subsequently produced in large numbers for the USAF, US Navy and US Coast Guard as a utility and SAR amphibian, the Albatross later found its way into the air arms of other countries. Today the small number of nations using the amphibian for SAR or maritime reconnaissance each have eight or fewer, including Greece, Mexico, Taiwan and Thailand. Mexican naval aircraft are of the HU-16A version, the original USAF model. The remainder are HU-16Bs. The HU-16Bs of the Hellenic Air Force, flown by the Navy, are among ex-Norwegian amphibians that had been produced in the early 1960s with anti-submarine capability, carrying search radar, MAD, a searchlight and depth charges as part of their equipment.

Number in crew: Four
Engines: Two 1425 h.p. Wright R-1820-76A radials
Length: 62 ft 10 in. (19.18 m)
Wing span: 96 ft 8 in. (29.46 m)
Gross weight: 37,500 lb (17,010 kg)
Maximum speed: 236 mph (379 km/h)

Grumman KA-6D (USA)
The KA-6D is a carrier-borne flight refuelling tanker, converted from A-6As; 62 were so modified, each capable of transferring a maximum of 21,000 lb (9500 kg) of fuel to other aircraft or up to 15,000 lb (6800 kg) at 250 nautical miles from the launch carrier.

Weapons: None, but can be used for day bombing if required

Grumman S-2 Tracker (USA)
The Tracker was developed for the US Navy as a carrier-borne anti-submarine aircraft, combining the tasks of 'hunter' and 'killer'. It first flew in 1952. It remained in US Navy first-line service until 1976, by which time updated variants had greatly improved its ASW capabilities. Conversion of some Trackers produced various utility aircraft for target towing and transport uses. Meanwhile, Trackers were acquired by other nations as land-based ASW aircraft and today S-2A, E, F and G versions remain operational for ASW, maritime reconnaissance and coastal patrol duties, together with similar CP-121s of the Canadian Armed Forces that are in reserve. Operating nations are Argentina, Brazil, South Korea, Peru, Taiwan, Thailand, Turkey, Uruguay and Venezuela.

Data: S-2E
Number in crew: Four
Engines: Two 1525 h.p. Wright R-1820-82WA radials
Length: 43 ft 6 in. (13.26 m)
Wing span: 72 ft 7 in. (22.13 m)
Gross weight: 29,150 lb (13,222 kg)
Maximum speed: 265 mph (426 km/h)
Ferry range: 1300 miles (2095 km)
Weapons: One bomb in bay and 60 echo-sounding depth charges, plus torpedoes, rockets or other weapons under the wings. Sonobuoys are carried in the nacelles

Hughes (McDonnell Douglas) Model 500MD Defender (USA)
The Model 500 Defender series covers military variants of the Model 500 civil helicopter for a variety of land and sea missions, ranging from anti-tank with TOW missiles to anti-submarine. The Taiwanese Navy operates 12 Model 500MD/ASW Defenders, flown from its ex-US Navy destroyers, while Spain has 11 basically similar aircraft which it uses as Z.13s from destroyers. Model 500MD/ASWs carry nose search radar and MAD. These are suited to ASW and anti-shipping roles.

Data: Model 500MD/ASW
Number in crew: Two
Engine: One 420 s.h.p. Allison 250-C20B turboshaft
Fuselage length: 24 ft 2½ in. (7.38 m)
Main rotor diameter: 26 ft 5 in. (8.05 m)
Gross weight: 3550 lb (1610 kg)
Cruising speed: 137 mph (221 km/h)
Combat radius, with 1¾ hours on station: 100 miles (160 km)
Weapons: Two torpedoes for ASW

Ilyushin Il-20 (USSR)
The Il-20 (NATO Coot-A) is an ELINT (electronics intelligence) and ECM derivative of the Il-18 turboprop airliner, seen by Western observers for the first time in 1978. It is thought that about 10 are in current Soviet use. Side-looking radar is housed in an underfuselage container and other sensor systems are carried.

Engines: Four 4250 e.h.p. Ivchenko AI-20M turboprops
Length: 117 ft 9 in. (35.90 m)
Wing span: 122 ft 9 in. (37.42 m)
Gross weight: 141,000 lb (64,000 kg)
Cruising speed: 419 mph (675 km/h)
Range: 4039 miles (6500 km)

Ilyushin Il-28 and Harbin H-5 (USSR/China)
First flown in 1948, the Il-28 became the first Soviet production jet bomber. Thousands were eventually built and put into the Soviet forces and exported, including a substantial number to China. Today Aviation of the People's Navy, the Chinese Navy air arm, retains approximately 50 as light bombers, while Chinese examples built at Harbin and designated H-5s also serve. It is thought that perhaps 100 H-5s are used by the Navy, each carrying one or two torpedoes, mines, bombs or depth charges up to a weight of 6614 lb (3000 kg). HZ-5 reconnaissance variants with electronic sensors or cameras in the weapons bay are also operated, and the Polish Naval Air Arm flies reconnaissance Il-28s.

Number in crew: Three
Engines: Two 5952 lb (2700 kg) non-afterburning
   Harbin Wopen-5 turbojets
Length: 57 ft 11 in. (17.65 m)
Wing span: 70 ft 4 in. (21.45 m)
Gross weight: 46,738 lb (21,200 kg)
Maximum speed: 560 mph (900 km/h)
Range: 1490 miles (2400 km)
Weapons: Two 23 mm cannon in nose and two in tail
      turret plus:
         Up to 6614 lb (3000 kg) of weapons, as
         detailed above

Ilyushin Il-38 (USSR)
See p. 69 for data.

Kaman SH-2F Seasprite (USA)
The SH-2F is the latest version of the Seasprite helicopter
that first joined the US Navy in the 1960s. To be found in
ones or twos on board many destroyers, cruisers and
frigates, it is known as a Mk I LAMPS helicopter (light
airborne multi-purpose system), suited to ASW and anti-
ship surveillance and targeting operations. Kaman
LAMPS helicopters will continue to operate alongside the
new Mk III LAMPS from Sikorsky. About 80 are active,
with others in reserve, each carrying Canadian Marconi
LN-66HP surveillance radar under the nose, MAD,
passive radiation detection receivers and other equipment. Up to 4000 lb (1814 kg) of cargo can be carried
externally.

*Kaman SH-2F Seasprite lands on board the frigate USS
O'Callahan for operations with the Pacific Fleet.*

Number in crew: Three
Engines: Two 1350 s.h.p. General Electric T58-GE-8F
   turboshafts
Length, blades folded: 52 ft 7 in. (16.03 m)
Main rotor diameter: 44 ft 0 in. (13.41 m)
Gross weight: 13,500 lb (6123 kg)
Maximum speed: 165 mph (265 km/h)
Range: 422 miles (679 km)
Weapons: One or two torpedoes,
      sonobuoys and marine markers

Kamov Ka-25 (USSR)
The Ka-25, known to NATO by the code name Hormone,
has been the standard Soviet ship-based anti-submarine
helicopter for many years, operating from battle cruisers,
cruisers, destroyers and frigates of the Soviet Navy as
well as the two Soviet helicopter carriers and the *Kiev*
class carriers. In addition to Hormone-A for ASW,
Hormone-Bs are to be found on board *Kiev* carriers and
cruisers/battle cruisers that are armed with long-range
cruise missiles, their job being to provide over-the-
horizon target acquisition and midcourse guidance for
the ship-launched missiles. A third version of the Ka-25 is
Hormone-C, an unarmed SAR and utility helicopter. As
for all Kamov helicopters, the Ka-25's power plant drives
two coaxial contra-rotating rotors, which does away with
the need for a tail rotor. Hormone-As are also flown by the
Indian Navy from destroyers (five aircraft), while Syria
has shore-based ASW aircraft (nine) and Yugoslavia flies
ten. Equipment for Hormone-A includes search radar in a
'chin' radome, dipping sonar, MAD towed 'bird' and
sonobuoys. Soviet Navy Hormone-A strength is some 120,
plus 25 Hormone-Bs.

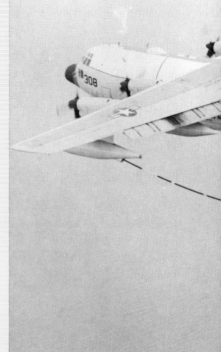

Data: Hormone-A
Number in crew: Four or five
Engines: Two 900 s.h.p. Glushenkov GTD-3F or two
  990 s.h.p. GTD-3BM turboshafts
Fuselage length: 32 ft 0 in. (9.75 m)
Rotor diameter (each): 51 ft 8 in. (15.74 m)
Gross weight: 16,500 lb (7500 kg)
Maximum speed: 136 mph (220 km/h)
Range: 250–400 miles (400–650 km)
Weapons: Torpedoes, nuclear depth charges, air-to-
        surface anti-shipping missiles and other
        stores

Kamov Ka-27 and Ka-32 (USSR)
These designations refer to the same basic helicopter,
known to NATO as Helix and currently superseding
Hormone with the Soviet Naval Air Force on-board ship.
Like the Ka-25, Helix has contra-rotating rotors and a
multi-fin/rudder tail unit but is far more powerful and has
a main cabin of increased size which opens up new
possibilities for Soviet ship-borne helicopters in assault
and VERTREP roles, while the helicopter's overall
dimensions allow its use of existing deck lifts and
hangars. Three versions have been identified to date,
Helix-A for ASW and corresponding with the designation Ka-27, Helix-B for the same roles as Hormone-B
and also a Ka-27, and a SAR/plane guard/VERTREP
model which probably shares the Ka-32 designation with
the civil model.

Data: Helix-A, estimated
Number in crew: Three
Engines: Two 2225 s.h.p. Isotov TV3-11 7V
  turboshafts

Soviet Navy Kamov Ka-25 Hormone-A
▲►
US Marine Corps Lockheed KC-130T tanker refuels two
USMC Sikorsky CH-53E Super Stallions.

Fuselage length: 36 ft 1 in. (11.00 m)
Rotor diameter (each): 54 ft 11 in. (16.75 m)
Maximum speed: 161 mph (260 km/h)
Mission radius: 186 miles (300 km)
Weapons: As for Hormone-A

Kawasaki P-2J (Japan)
The Lockheed P-2 Neptune land-based anti-submarine
aircraft first flew in 1945 and was subsequently put into
service by the US Navy and the air arms of other
countries. In US service it was later superseded by the P-3
Orion, and those in foreign use also disappeared as newer
aircraft became available (such as the Atlantic). The
Neptune in P-2H form had been built in Japan for the
JMSDF by Kawasaki, which went on to develop a new
variant as the P-2J. Longer, better equipped and using
turboprop engines in place of piston radials, this model
entered JMSDF service in 1969 and today 58 are operated
alongside a much smaller number of PS-1s and P-3Cs.
Detection equipment includes AN/APN-187B-N Doppler
radar, AN/APS-80-N search radar, MAD and sonobuoys.

Number in crew: 12
Engines: Four 2850 e.h.p. General Electric T64-IHI-10
  turboprops, and two 3085 lb (1400 kg) thrust
  Ishikawajima J3-IHI-7C turbojets
Length: 95 ft 10¾ in. (29.23 m)

Wing span: 97 ft 8½ in. (29.78 m)
Gross weight: 75,000 lb (34,019 kg)
Cruising speed: 250 mph (402 km/h)
Range: 2765 miles (4450 km)
Weapons: Up to 8000 lb (3630 kg) of torpedoes, depth
         charges and bombs carried in the bay, and
         rockets under the wings

Lockheed C-130 Hercules (USA)
Stalwart medium transport aircraft of the West and
others, the Hercules is in worldwide service with some 55
nations. In a naval context, the Hercules is used as an
assault transport and in-flight refuelling tanker (with
probe and drogue gear) by the US Marine Corps and
Reserve, a communications relay aircraft by the US Navy,
an SAR aircraft by the USAF for the recovery of re-
entered space capsules and rescue of aircrew, and as a
transport/SAR and maritime patrol aircraft by Portugal,
Indonesia and Malaysia.

Data: C-130H
Number in crew: C-130H-MP (maritime patrol) has a
    flight crew of four, plus specialists operating sea
    search radar, etc.
Engines: Four 4508 e.h.p. Allison T56-A-15 turboprops
Length: 97 ft 9 in. (29.79 m)
Wing span: 132 ft 7 in. (40.41 m)
Gross weight: 155,000 lb (70,310 kg)
Cruising speed: 374 mph (602 km/h)
Mission radius, with 2½ hours on station search time:
    2070 miles (3333 km)

*The maximum payload of the C-130H-MP is 41,074 lb
(18,630 kg)

Lockheed F-104G Starfighter (USA/West Germany)
The Starfighter was designed for the USAF as one of the
'Century Series' fighters of the 1950s, its Mach 2
performance, slim pointed fuselage and short-span wings
giving rise to the expression 'Manned Missile'. In the
event it found much greater favour with European air
forces and European production of the Starfighter
included the multi-mission F-104G. Although being
superseded by the Tornado in a naval strike role, the
German Marineflieger still has F-104Gs for this role and
for reconnaissance and training (TF-104G), carrying
Kormoran air-to-surface anti-shipping missiles.

Number in crew: One
Engine: One 15,800 lb (7167 kg) thrust with
    afterburning General Electric J79-GE-11A turbojet
Length: 54 ft 9 in. (16.69 m)
Wing span: 21 ft 11 in. (6.68 m)
Gross weight: 25,027 lb (11,352 kg)
Maximum speed: Mach 2
Weapons: Kormoran air-to-surface anti-shipping
    missile

Lockheed L-188E Electra (USA)
Three Electras are flown by the Argentine Navy on
maritime reconnaissance missions, and a further four are
used as transports.

Engines: Four 3750 e.h.p. Allison 501-D13A turboprops

Lockheed P-3 Orion (USA)

The standard shore-based anti-submarine and maritime reconnaissance aircraft of the US Navy, and also used in P-3A, B and C versions by Australia, Canada (as the CP-140 Aurora), Iran, Japan (as the Kawasaki-Lockheed P-3C), the Netherlands, New Zealand, Norway and Spain, the Orion is based on the airframe of the Electra turboprop airliner. In 1985, the US Navy and Reserve operated 368 Orions, early production P-3As and a number of the more powerful P-3Bs with the Reserve and the active US Navy using mostly P-3Cs in its first-line strength of 218, plus 40 P-3B/Cs with operational conversion units. Twelve EP-3E Orions serve with a Navy ELINT squadron. Like the P-3B, the latest P-3C is powered by 4910 e.h.p. Allison T56-A-14 turboprops, but has updated avionics, mainly for improved detection capability, and also has the ability to launch Harpoon. Like most highly sophisticated ASW aircraft, it can detect by electronic, sonic and magnetic means, MAD being carried in a tail 'sting'.

Data: P-3C
Number in crew: Ten
Engines: See above
Length: 116 ft 10 in. (35.61 m)
Wing span: 99 ft 8 in. (30.37 m)
Gross weight: 135,000 lb (61,235 kg)
Maximum speed: 473 mph (761 km/h)
Mission radius: 2383 miles (3835 km)
Weapons: Weapons bay, or a combination of:
    One 2000 lb mine
    Three 1000 lb mines
    Three to eight depth bombs
    Two nuclear depth bombs

Eight torpedoes
Underwing weapons:
    Mines, rockets or Harpoon air-to-surface
    anti-shipping missiles

Lockheed S-3A Viking (USA)

The Viking is the standard US Navy carrier-borne anti-submarine fixed-wing aircraft, 110 currently serving with carrier squadrons and others with an operational conversion unit. First flown in 1972, production of 187 Vikings ended in 1978, with initial deployment at sea on board USS *John F. Kennedy* taking place in 1975. Detection equipment includes AN/APS-116 search radar, AN/ASQ-81 MAD, a forward-looking infra-red scanner (FLIR), ECM receiving equipment, and sonobuoy receiver. Under an improvement programme, plans have been put forward to modify 160 existing S-3As to S-3B standard and perhaps construct new S-3Bs, each with increased acoustic processing capability, greater ESM (electronic support measure) capability, and the ability to launch Harpoon among updates. The Viking is unusual on the carrier deck for having two podded engines carried under the high wings and, like several Grumman carrier aircraft, has side-by-side seating for the flight crew.

Number in crew: Four
Engines: Two 9275 lb (4207 kg) thrust General Electric
    TF34-GE-2 turbofans
Length: 53 ft 4 in. (16.26 m)
Wing span: 68 ft 8 in. (20.93 m)
Gross weight: 42,500 lb (19,277 kg)
Maximum speed: 518 mph (834 km/h)
Range: more than 2300 miles (3700 km)
Weapons: Split weapons bay for:
    Four torpedoes, depth bombs, destructors,
    mines or bombs
    Underwing pylons for:
    Bombs, mines, destructors, flares, rockets or
    air-to-surface missiles (will carry
    Harpoon)

An S-3A Viking leaves the flight deck of USS John F. Kennedy. *Note the compression in the starboard oleo, which is still in contact with the deck.*

McDonnell Douglas A-4 Skyhawk (USA)

First flown as a prototype in 1954, the Skyhawk was once a familiar sight on-board US Navy aircraft carriers. Nevertheless, although it is being superseded by Hornets and Harrier IIs, it remains an important part of the US Marine Corps' attack strength in A-4M form and as a forward air control aircraft as the OA-4M. It is also in widespread use with USMC reserve squadrons and continues in first-line service with several foreign air forces and navies, including Argentina whose naval A-4Qs saw action during the Falklands conflict. The Skyhawk is distinctive by its low-mounted delta-type wings.

Data: A-4M Skyhawk II
Number in crew: One
Engine: One 11,200 lb (5080 kg) Pratt & Whitney
   J52-P-408A turbojet
Length: 40 ft 3¼ in. (12.27 m)
Wing span: 27 ft 6 in. (8.38 m)
Gross weight: 24,500 lb (11,113 kg)
Maximum speed with warload: 645 mph (1038 km/h)
Range: 2055 miles (3307 km) with maximum auxiliary
   fuel
Weapons: Two 20 mm Mk 12 cannon in wings plus:
         Wide variety of bombs, missiles and other
         weapons to .10,000 lb (4535 kg)

McDonnell Douglas F-4 Phantom II (USA)

In 1960 the first Mach 2 F-4s were received by the US Navy, the start of a long partnership that eventually raised the fighting capability of the US Navy and Marine Corps to unprecedented heights. Although designed as a naval missile fighter, it was also taken into service by the USAF and widely exported, the Royal Navy taking the F-4 to sea. Today, with the US carrier fleet converted to Tomcats and the Royal Navy deploying 'Harrier carriers', the F-4 is no longer carrier borne. However, although it is slowly being replaced by the Hornet within the US Marine Corps, the F-4N and F-4S remain important all-weather fighter-bombers with this service and in the Navy Reserve and Marine Corps Reserve, in addition to those still in Navy markings but on land. In 1985 about 228 US naval F-4s remained active, with others in store, the largest number with fighter squadrons of the USMC.

Number in crew: Two
Engines: Two 17,000 lb (7710 kg) thrust General
   Electric J79-GE-8 turbojets
Length: 58 ft 3 in. (17.76 m)
Wing span: 38 ft 7½ in. (11.70 m)
Gross weight: 54,600 lb (24,765 kg)
Maximum speed: above Mach 2
Combat radius: 900 miles (1450 km)
Weapons: Sparrow and Sidewinder missiles, up to
         16,000 lb (7250 kg) of bombs or other
         weapons

McDonnell Douglas F/A-18A Hornet (USA)

See p. 100 for data.

McDonnell Douglas/British Aerospace Harrier II (USA/UK)

First becoming operational with the US Marine Corps in 1985, the Harrier II has been designed as the follow-on V/STOL close support aircraft to the world-famous British Harrier. All three nations operating the original Harrier are receiving Harrier IIs. Designed to use the Pegasus vectored-thrust engine of the Harrier but to gain improvements in weapon-carrying capability and range through airframe changes, the Harrier II has new supercritical wings that are more than 14% larger in area than those of the original Harrier. These are of increased span and are constructed of carbonfibre and other composite materials. Sweepback has been reduced on the leading edges but rounded root extensions have been added. The USMC is expected to receive some 328 Harrier IIs by the early years of the next decade, under the designation AV-8B, while the RAF has plans for 60 as Harrier GR.5s. Spain is to receive 12 AV-8Bs, most of which will be used on its new aircraft carrier. Marine Corps AV-8Bs are intended to replace existing Harriers with three operational and one training squadrons and five other squadrons at present flying Skyhawks.

Number in crew: One (two in TAV-8B)
Engine: One 21,500 lb (9750 kg) thrust Rolls-Royce
   F402-RR-406 turbofan in AV-8B; 21,750 lb (9865 kg)
   thrust Pegasus Mk 105 in GR.5)
Length: 46 ft 4 in. (14.12 m)
Wing span: 30 ft 4 in. (9.25 m)
Gross weight: 29,750 lb (13,494 kg)
Maximum speed: Mach 1.1
Combat radius with one-hour loiter, weapons and no
   auxiliary fuel: 172 miles (278 km)
Weapons: One 25 mm GAU-12/U five-barrel cannon
         (AV-8B) or two 25 mm cannon (GR.5)
         plus:
         Two or four Sidewinder or Magic air-to-air
         missiles
         Maverick air-to-surface missiles
         Up to 16,500 lb (7484 kg) of bombs, laser
         guided bombs, rockets, etc.

Mil Mi-8 (USSR)

Known to NATO by the code name Hip, this large helicopter was first seen in 1961 and is in widespread civil and military use. Military versions include very heavily armed assault and airborne communications models, but in a naval context it is used by the Soviet Naval Air Force (with the Mi-6/NATO Hook) as a transport and training helicopter, while the Polish Navy operates five and the Yugoslav Navy has 20 for transport and coastal patrol (which it may operate alongside a number of SOKO-built Anglo-French Aérospatiale SA 342 Gazelles known locally as Partizans).

Number in crew: Two or three. Up to 32 passengers,
   8820 lb (4000 kg) of cargo or stretchers
Engines: Two 1700 s.h.p. Isotov TV2-117A turboshafts
Fuselage length: 59 ft 7 in. (18.17 m)
Main rotor diameter: 69 ft 10 in. (21.29 m)
Gross weight: 26,455 lb (12,000 kg)
Maximum speed: 161 mph (260 km/h)

Myasishchev M-4 Bison-C maritime reconnaissance aircraft.

Mil Mi-14 (USSR)

Known to NATO by the code name Haze, this is one of the most important helicopters of the Soviet Naval Air Force, which has about 100 in service for land-based ASW (Haze-A) and 10 for mine-countermeasures (Haze-B). Based on the Mi-8 airframe but with a boat hull, sponsons and an under-tail float to make it amphibious, it carries search radar in a 'chin' radome, a MAD towed 'bird' and other equipment for the ASW role, with weapons housed in an under-hull bay. ASW Haze-As are also operated by Bulgaria, Cuba, East Germany, Libya and Poland.

Number in crew: Four or five
Engines: Probably two 2200 s.h.p. Isotov TV3-117
    turboshafts
Length, main rotor diameter and gross weight: Similar
    to Mi-8
Weapons: Torpedoes and depth charges

Myasishchev M-4 (USSR)

This huge turbojet bomber was designed as a stablemate of the Tupolev Tu-95 turboprop bomber, but was nothing like as successful because of disappointing range and service ceiling. However, some 45 remain assigned the role of long-range strategic bomber (NATO code name Bison-As) with the Soviet Air Force, while some 30 other Bison-As are used as in-flight refuelling tankers for Soviet bombers. In a naval context, a limited number of Bison-Bs and improved Bison-Cs are in service in maritime reconnaissance roles, although they are far less important than the Tu-142s, Tu-16s and Tu-22Ms flown in maritime operations.

Engines: Four 19,180 lb (8700 kg) thrust Mikulin
    AM-3D turbojets
Length: 154 ft 10 in. (47.20 m)
Wing span: 165 ft 7 in. (50.48 m)
Gross weight: 350,000 lb (158,750 kg)
Maximum speed: 620 mph (998 km/h)
Range: about 4970 miles (8000 km)
Weapons: Ten 23 mm cannon plus:
              Bombs

Panavia Tornado (UK/West Germany/Italy)

The Tornado has been entering service in two major forms, in the IDS interdictor/strike model for all the air forces of the participating nations and as the ADV (air defence variant) interceptor for the RAF. Of the IDS Tornadoes delivered or to be completed, 112 are destined for the German Navy, whose Marinefliegergeschwader 1 and 2 are replacing F-104G, TF-104G and RF-104G Starfighters in the roles of anti-shipping and coastal target strike and reconnaissance. In configuration, the Tornado is a variable-geometry aircraft with an extremely tall single tailfin. Although the German Navy Tornadoes have a specific maritime role, all IDS aircraft can carry anti-shipping weapons if required.

Number in crew: Two
Engines: Two 16,000 lb (7257 kg) thrust with
    afterburning Turbo-Union RB199-34R Mk 101
    turbofans on early aircraft, superseded by more
    powerful Mk 103 engines
Length: 54 ft 10$\frac{1}{4}$ in. (16.72 m)
Wing span: Spread 45 ft 7$\frac{1}{2}$ in. (13.91 m)
             Swept 28 ft 2$\frac{1}{2}$ in. (8.60 m)
Gross weight: about 60,000 lb (27,215 kg)
Maximum speed: Mach 2.2
Combat radius: 863 miles (1390 km)
Weapons: Two 27 mm IWKA-Mauser cannon in
            fuselage plus:
            Kormoran or Sea Eagle anti-shipping
               missiles or a variety of other weapons up
               to about 19,840 lb (9000 kg)

Pilatus Britten-Norman ASW Maritime Defender (UK)

Based upon the successful turbine Islander feederline transport, the Maritime Defender is a maritime patrol aircraft that can be armed with torpedoes and other weapons. It carries a 360° radar scanner, and other equipment to fit requirements can include FLIR, MAD and sonobuoys. The Indian Navy has 18 Maritime Islanders, used for maritime patrol and communications.

Number in crew: Three
Engines: Two 400 s.h.p. Allison 250-B17C turboprops
Length: 36 ft 3$\frac{3}{4}$ in. (11.07 m)
Wing span: 53 ft 0 in. (16.15 m)

Cruising speed: 195 mph (315 km/h)
Mission radius: 115 miles (185 km)
Weapons: Two lightweight torpedoes or depth
          charges, guns, rockets or other weapons

*An AEW (airborne early warning) version of the Maritime Defender has also been developed.

Rockwell International OV-10 Bronco (USA)

First flown in 1965, the Bronco twin-boom counter-insurgency aircraft was taken into USAF and US Marine Corps service for forward air control, observation and ground support missions, and was exported. A number of USMC aircraft were loaned to the US Navy for duties in Vietnam which included helicopter escort, the Bronco's slow speed but good weapon load making it ideal as a temporary measure for this role. US Marine Corps aircraft were designated OV-10A and about 37 remain in active and reserve service, plus 17 similar aircraft modified to OV-10D standard for night observation and surveillance (NOS) and featuring more powerful engines, additional weapons (including a cannon carried under the fuselage), FLIR sensor and laser target designator in a nose turret, chaff/flare dispensers and other equipment.

Data: OV-1OD
Engines: Two 1040 e.h.p. Garrett T76-G-420/421
   turboprops
Length: 44 ft 0 in. (13.41 m)
Wing span: 40 ft 0 in. (12.19 m)
Gross weight: 14,444 lb (6552 kg)
Maximum speed: 288 mph (463 km/h)
Mission radius: 228 miles (367 km)
Weapons: One 20 mm General Electric M97 cannon in
          underfuselage turret plus:
          Up to 3600 lb (1633 kg) of bombs, rocket
          and guns.
          Provision for two Sidewinder air-to-air
          missiles for self-defence

Saab SH-37 Viggen (Sweden)

The Viggen multi-mission fighter first flew as a prototype in 1967 and in 1971 the Swedish Air Force began receiving initial production aircraft in the form of AJ 37 attack aircraft. Subsequent versions have included the JA 37 interceptor, SF 37 armed photographic reconnaissance aircraft, SK 37 two-seat trainer and the SH 37, the latter an all-weather maritime reconnaissance version with attack capabilities. Production deliveries of the SH 37 began in 1975. The SH 37 carries surveillance radar in the nose plus, optionally, ECM pods, night reconnaissance and camera pods.

Number in crew: One
Engine: One 26,015 lb (11,800 kg) thrust with
   afterburning Volvo Flygmotor RM8A turbofan
Length: 53 ft 5¾ in. (16.30 m)
Wing span (main wings): 35 ft 9 in. (10.60 m)
Span of foreplanes: 17 ft 10½ in. (5.45 m)
Gross weight: about 33,070 lb (15,000 kg)
Maximum speed: more than Mach 2
Mission radius: about 310–620 miles (500–1000 km)
Weapons: Two Sidewinders can be carried for
          defence.
          Attack weapons if required

Shenyang J-5 and Mikoyan-Gurevich MiG-17 (China/USSR)

J-5 is the Chinese designation of the Soviet Mikoyan-Gurevich MiG-17 day or limited all-weather fighter and attack aircraft, formerly built in China for the Air Force of the People's Liberation Army and Aviation of the People's Navy, and exported. Chinese-built aircraft began to come off production lines at the end of the 1950s. A reasonable number remain in Navy use. MiG-17 types are known to NATO as Fresco, and the Polish Naval Air Arm operates about 39, mostly as strike aircraft used by three squadrons but a handful also in a reconnaissance role.

Data: Based upon MiG-17F
Number in crew: One
Engine: One 7450 lb (3380 kg) thrust with
   afterburning Klimov VK-1A turbojet
Length: 37 ft 3 in. (11.36 m)
Wing span: 31 ft 6 in. (9.60 m)
Gross weight: 13,380 lb (6070 kg)
Maximum speed: 711 mph (1145 km/h)
Range with auxiliary fuel tanks, as a fighter-bomber:
   870 miles (1400 km)
Weapons: One 37 mm N-37 or two 23 mm NR-23
          cannon in the nose plus:
          Up to 1102 lb (500 kg) of bombs or rockets

Shenyang/Tianjin J-6 (China)

J-6 is the Chinese designation of the Mikoyan MiG-19 fighter (NATO code name Farmer), built in China for the Air Force of the People's Liberation Army and Aviation of the People's Navy, and exported. Although the design originates from the Soviet Union in the early 1950s, the J-6 has been updated and remains the most important fighter and attack aircraft of the Chinese forces and production probably continues at both Shenyang and Tianjin. J-6s of various versions form part of the Navy's strength of some 600 fighter-bombers, all land based. These provide air defence protection to naval bases, form part of the nation's main air defence system, and undertake strike and possible anti-shipping roles.

Number in crew: One
Engines: Two 7165 lb (3250 kg) thrust with
   afterburning Shenyang Wopen-6 turbojets
Length: 41 ft 4 in. (12.60 m)
Wing span: 30 ft 2 in. (9.20 m)
Gross weight: estimated at 22,050 lb (10,000 kg)
Maximum speed: Mach 1.45
Range with auxiliary fuel tanks: 1366 miles (2200 km)
Weapons: Two/three 30 mm NR-30 cannon plus:
          Two or four air-to-air missiles, rockets or
          bombs

Shin Meiwa PS-1 and US-1 (Japan)

In 1967 Shin Meiwa flew the prototype of a new ASW aircraft for the JMSDF, unusual in being a large flying-boat. Fourteen are currently used by the 31st Air Group for ASW, each carrying AN/APS-80-N search radar, AN/APN-187C-N Doppler radar, MAD, dipping sonar, sonobuoys and much else. In 1975 the JMSDF began receiving a new version of the flying-boat, this time an air/sea rescue amphibian as the US-1. Ten have been built, the latest examples with 3490 s.h.p. T64-IHI-10J turboprops and known as US-1As. These can accommodate 20 passengers or 12 stretchers and equipment includes the same search and Doppler radars as used by

◁ ▲

US Marine Corps Rockwell International OV-10D Bronco with NOS equipment.

◁ ▼

German Marineflieger Panavia Tornado variable-geometry combat aircraft.

▲

One of the few large flying-boats in military service today is the Shin Meiwa PS-1, flown by the JMSDF on ASW missions.

the PS-1, marine markers, air-droppable message cylinders, floating lights, two life rafts in containers, three lifebuoys, a powered lifeboat and much else.

Data: PS-1
Number in crew: Ten
Engines: Four 3060 e.h.p. Ishikawajima/General
    Electric T64-IHI-10 turboprops
Length: 109 ft 9 in. (33.46 m)
Wing span: 108 ft 9 in. (33.15 m)
Gross weight: 94,800 lb (43,000 kg)
Maximum speed: 340 mph (547 km/h)
Range: 2614 miles (4207 km)
Weapons: Four 330 lb (150 kg) anti-submarine
    bombs and smoke bombs in bay,
    plus four torpedoes in underwing
    pods and 5 in. rockets

Sikorsky CH-53A, CH-53D and RH-53D Sea Stallion (USA)

The Sea Stallion entered US Marine Corps service from 1966 as a heavy assault helicopter, able to transport 37 marines or combat equipment such as a 105 mm howitzer or two jeeps. In 1968 a CH-53A demonstrated its ability to be rolled and looped and a similar CH-53A also performed the first automatic terrain clearance flight by a helicopter. From 1969 an improved version began equipping the USMC, the CH-53D allowing up to 55 troops to be carried, or 24 stretchers (as for the CH-54A) or cargo. Both versions remain in service, loading via a rear ramp under the tailboom. The total number of Sea Stallions built for the USMC was 265, many of which remain active and with one reserve squadron. In the early 1970s the US Navy was loaned 15 CH-53As with which to establish its first helicopter mine-countermeasures squadron as HM-12. From 1973 this squadron began receiving purpose-built RH-53Ds for this new role, each with winches for streaming mechanical, acoustic and magnetic mine-countermeasures equipment and armed with two 0.50-in. machine-guns to blast surface mines. In 1985 the US Navy had 14 RH-53Ds in two MCM squadrons, although initial deliveries were beginning of the improved MH-53E. It was RH-53s from USS *Nimitz* that were used in the tragic 'Operation Evening Light' mission of April 1980, which attempted to release US nationals from Iran. The Iranian Navy itself had previously received six similar RH-53Ds, of which two are thought to remain in use.

Data: CH-53D
Number in crew: Three
Engines: Two 3925 s.h.p. General Electric T64-GE-413
    turboshafts
Fuselage length: 67 ft 2 in. (20.47 m)

Main rotor diameter: 72 ft 3 in. (22.02 m)
Gross weight: 42,000 lb (19.050 kg)
Maximum speed: 196 mph (315 km/h)
Range: 257 miles (413 km)

Sikorsky CH-53E Super Stallion and MH-53E Sea Dragon
(USA)
To improve even further upon the lifting capability of the
US Marine Corps' heavy assault Sea Stallion, the CH-53E
was developed with three engines, uprated transmission,
a seven (instead of six) blade main rotor, longer fuselage
and other changes, the resulting helicopter representing
the largest and most powerful built anywhere other than
in the USSR. First flown in 1974, full production CH-53Es
were first received by the USMC in mid-1981 and within
three years more than 70 were in service with the USMC
and USN. Perhaps 200 or more could be acquired by the
1990s, although not as many as that have been ordered
yet. The CH-53E has a watertight fuselage for its
amphibious operations and, like some other Navy
helicopters, it can take on fuel in in-flight refuelling
operations from tanker aircraft (such as the Hercules) or
from ships while hovering. Like the CH-53D, it can
accommodate 55 troops, but its cargo-lifting capacity is
much improved over the CH-53D and, in the role to rescue
disabled aircraft or lift damaged aircraft from carrier
decks, it can lift external loads of up to 33,685 lb (15,279
kg). Deliveries are now also beginning of an advanced
mine-countermeasures variant of the Super Stallion
designated MH-53E Sea Dragon (see RH-53D), with an
endurance of four hours.

Data: CH-53E
Number in crew: Three
Engines: Three 3696 s.h.p. General Electric T64-GE-
    416 turboshafts
Fuselage length: 73 ft 4 in. (22.35 m)
Main rotor diameter: 79 ft 0 in. (24.08 m)
Gross weight: 73,500 lb (33,339 kg)
Maximum speed: 196 mph (315 km/h)
Ferry range: 1290 miles (2075 km)

Sikorsky SH-60B Seahawk (USA)
To complement existing Seasprite LAMPS helicopters
serving with the US Navy, the SH-60B Seahawk was
developed as an improved Mk III LAMPS based upon the
airframe of the Army's Black Hawk assault helicopter.
Primarily for ASW and ASST (anti-ship surveillance and
targeting) but with secondary roles including VERTREP
and SAR, its deployment began in 1984 with warships of
the US Navy. The USN is expected to receive 204. In
addition, eight similar S-70B-2RAWS are going to the
Royal Australian Navy for service from 1987 on board
four FFG 7 class frigates, and the JMSDF will eventually
use the helicopter as a Sea King replacement. US Navy
Seahawks carry Texas Instruments AN/APS-124 search
radar in a circular radome under the forward fuselage,
MAD, ESM, sonobuoys and much else, and have more
powerful engines and greater fuel capacity than the
Army's Black Hawks.

Number in crew: Three
Engines: Two 1690 s.h.p. General Electric T700-GE-401
    turboshafts
Fuselage length: 50 ft 0¾ in. (15.25 m)
Main rotor diameter: 53 ft 8 in. (16.36 m)
Gross weight: 21,884 lb (9926 kg)
Maximum speed: 145 mph (234 km/h)
Weapons: Two torpedoes or other weapons

Sukhoi Su-17 (USSR)
The Soviet Naval Air Force reportedly operates some 65
Su-17 attack aircraft in an anti-shipping role, these
aircraft known to NATO as Fitter-Cs. The Su-17 is a high-
performance combat aircraft, with an unusual wing
arrangement comprising a swept fixed centre-section and
variable-geometry outer panels that can sweep at angles
estimated to be between 28° and 62°.

Number in crew: One
Engine: One 24,690 lb (11,200 kg) thrust with
    afterburning Lyulka AL-21 F-3 turbojet
Fuselage length: 61 ft 6 in. (18.75 m)
Wing span: Spread 45 ft 11 in. (14.00 m)
            Swept 34 ft 9 in. (10.60 m)
Gross weight: 39,020 lb (17,700 kg)
Maximum speed: above Mach 2
Combat radius: 224–390 miles (360–630 km)
Weapons: Two 30 mm NR-30 cannon plus:
            6615 kg (3000 kg) of nuclear or
            conventional bombs, air-to-surface
            missiles (including AS-7 Kerry with a 100
            kg high-explosive warhead), rockets and
            other stores

Tupolev Tu-16 and Xian H-6 (USSR/China)
See p. 84 for data.

Tupolev Tu-22 (USSR)
The Soviet Naval Air Force is thought to operate 35
examples of the Tu-22 supersonic bomber in maritime
reconnaissance and ECM roles, the cameras of some
Blinder-Cs (NATO name) being supplemented by
electronic equipment suggesting ECM (but possibly
directed towards intelligence). The Tu-22 is unusual in
having its two large turbojet engines mounted each side
of the vertical tail.

Data: Estimated
Number in crew: Three
Engines: Two 30,865 lb (14,000 kg) thrust Koliesov
    VD-7 turbojets
Length: 133 ft 0 in. (40.53 m)
Wing span: 90 ft 10 in. (27.70 m)
Gross weight: 185,000 lb (83,900 kg)
Maximum speed: Mach 1.4
Mission radius: 1925 miles (3100 km)
Weapons: One 23 mm NR-23 cannon in tail

Tupolev Tu-22M (USSR)
See p. 84 for data.

Tupolev Tu-142 (USSR)
Tu-142 is the designation given to Soviet Naval Air Force
examples of the Tu-95 long-range turboprop bomber
known to NATO as Bear. It is believed that some 95 are
used in a naval context comprising Bear-Ds with an
undernose radar scanner, underfuselage radome and
other equipment for maritime reconnaissance and target
acquisition for long-range missiles, Bear-E maritime
reconnaissance aircraft with cameras and other equipment, and Bear-Fs for anti-submarine duties and some
carrying MAD in addition to other detection equipment.

Soviet Tupolev Tu-22 Blinder photographed by an
interceptor of the Royal Danish Air Force.

▲
The French Navy employs the only examples of the Vought Crusader fighter still at sea.

▶
Vought TA-7C (foreground) version of the subsonic Corsair II attack aircraft flies alongside a single-seater.

Though the basic Tu-95/Tu-142 design dates from the 1950s, the capability and importance of the bomber are such that limited production has continued to maintain the number in operational service. The choice of turboprops was interesting, the original concept of using four of the world's most powerful, instead of turbojets as used in other Soviet bombers and by the equivalent US B-52 to result in excellent fuel economy and thereby long range, proved sound. Naval Bears are most importantly flown over the Atlantic and North Sea, and operate from Europe, Cuba, Angola and Vietnam.

Engines: Four 14,795 e.h.p. Kuznetsov NK-12MV
    turboprops
Length: 162 ft 5 in. (49.50 m)
Wing span: 167 ft 8 in. (51.10 m)
Gross weight: 414,470 lb (188,000 kg)
Maximum speed: 575 mph (925 km/h)
Mission radius: 5150 miles (8285 km)
Weapons: Cannon in tail position plus:
            Air-to-surface anti-shipping missiles or
            bombs

Vought A-7E and TA-7C Corsair II (USA)
The Corsair II was designed as a subsonic, single-seat, carrier-based attack aircraft for the US Navy, able to carry a heavier warload than the then current A-4E version of Skyhawk. To keep development costs down, the Corsair II was based partially on the in-service F-8 Crusader fighter, although its high-mounted wing was of fixed incidence. First flying in 1965, the Corsair II eventually went into Navy service and was also adopted

by the USAF. Today, while the Hornet is gradually superseding it on the carrier deck, the US Navy and Reserve still have many in service, the active USN operating A-7Es and similar TA-7C two-seat trainers. Newly built A-7Es, with forward-looking infra-red sensor (FLIR) equipment, were delivered from 1978 and currently most of the Navy's A-7Es assigned to active attack squadrons are FLIR equipped, accounting for well over 200 aircraft. The TA-7Cs represent conversions of A-7Bs and Cs into two-seaters, the programme aimed mainly at providing an A-7 for actual carrier training operations, while also improving the aircraft's combat capability.

Data: A-7E FLIR
Engine: one 15,000 lb (6803 kg) Allison TF41-A-2
    turbofan
Length: 46 ft 1½ in. (14.06 m)
Wing span: 38 ft 9 in. (11.80 m)
Gross weight: 42,000 lb (19,050 kg)
Maximum speed: 691 mph (1112 km/h)
Ferry range: up to 2861 miles (4605 km)
Weapons: One 20 mm M61A1 six-barrel gun plus:
        Up to 15,000 lb (6805 kg) of bombs, rockets,
        air-to-air and air-to-surface missiles

Vought F-8 Crusader (USA)
The Crusader was put into production for the US Navy as a carrier-borne supersonic fighter. Although initially only suited to daytime operations, it featured a unique variable-incidence wing to ease carrier operations. As designed, the wing incidence could be increased before take-off and during the landing approach to give good low-speed flying characteristics while keeping the fuselage nose down for improved pilot visibility. Having completed aircraft carrier qualification trials on board USS *Forrestal* in April 1956 and establishing a new US speed record the same year, the F-8A initial version first entered service on board USS *Saratoga* in 1957. Subsequent versions offered more engine power and improved all-weather capabilities. The Crusader was withdrawn from first-line service as a fighter with the US Navy during the 1970s, but the French Navy still operates F-8E(FN) all-weather fighters and has some 14 on active strength in two flights. One Crusader flight is carried on board *Foch*.

Number in crew: One
Engine: One 18,000 lb (8165 kg) thrust with
    afterburning Pratt & Whitney J57-P-20 turbojet
Length: 54 ft 6 in. (16.61 m)
Wing span: 35 ft 8 in. (10.87 m)
Gross weight: 34,000 lb (15,420 kg)
Maximum speed: approaching Mach 2
Combat radius: 600 miles (965 km)
Weapons: Four cannon in nose plus:
        Air-to-air missiles of Matra R 530, R 550
        Magic and Sidewinder types
Naval operators: France

Westland Lynx (UK)
The fairly small Lynx multi-purpose helicopter is an Anglo-French type, Westland co-operating in its development and production with Aérospatiale. In its naval forms, the Lynx is to be found on-board ship and shore based. It has been used as a Wasp replacement by the Royal Navy on-board destroyers and frigates and by the French Navy as the air element on board destroyers and the helicopter carrier *Jeanne d'Arc*. At sea its main task is that of the anti-submarine and anti-surface vessel roles, Royal Navy helicopters carrying Ferranti Seaspray search and tracking radar in the nose and with provision for dipping sonar. MAD has also been used. Sea Skua anti-ship missiles are among the weapon options of these Lynx helicopters. French Navy ASW/ASV Lynx helicopters carry Omera-Seguid ORB-31-W radar, dipping sonar and can carry AS.12 wire-guided air-to-surface missiles. Apart from these navies, Lynx is flown by the navies of Argentina (ASW role), Brazil (ASW), Denmark (ASW and maritime patrol), West Germany (ASW), the Netherlands (ASW and SAR), and Nigeria (ASW, SAR and maritime patrol), plus Norway (Air Force for SAR).

Data: Royal Navy HAS.3
Number in crew: Two or more
Engines: Two 1120 s.h.p. Rolls-Royce Gem 41-1
    turboshafts
Fuselage length: 39 ft 1½ in. (11.92 m)
Main rotor diameter: 42 ft 0 in. (12.80 m)
Gross weight: 10,500 lb (4763 kg)
Cruising speed: 144 mph (232 km/h)
Range: 368 miles (593 km)
Weapons: Two torpedoes (including Sting Ray)
        Two depth charges
        Four Sea Skua air-to-surface anti-shipping
        missiles
        Four AS.12 air-to-surface missiles

Westland Sea King (UK)
See p. 164.

Westland Wasp (UK)
The Wasp first flew in 1958 and joined the Royal Navy as a standard frigate-borne anti-submarine helicopter. Because of its small size, it is unable to carry its own submarine detection equipment but instead performs as the 'killer' while directed by the ship's equipment. With the advent of the Lynx, the Wasp HAS.1 is gradually being phased out from Royal Navy frigates. Also capable of search and rescue, utility and training missions, the Wasp is still active in an ASW role with Indonesia (10), New Zealand (7) and South Africa (10), and is used for SAR and utility in Brazil (8).

Number in crew: Two (can accommodate three
    passengers or a stretcher)
Engine: One derated 710 s.h.p. Rolls-Royce/Bristol
    Nimbus 503 turboshaft
Fuselage length: 30 ft 4 in. (9.24 m)
Main rotor diameter: 32 ft 3 in. (9.83 m)
Gross weight: 5500 lb (2495 kg)
Maximum speed: 120 mph (193 km/h)
Range: 303 miles (488 km)
Weapons: Two torpedoes or other weapons

Yakovlev Yak-38 (USSR)
See p. 49 for data.

Index

Acknowledgements

Artwork

Will Stephen 73, 80, 94, 98, 101, 102, 104–5, 108–9, 120, 128–9.

Photographs

Aérospatiale 63, 126 top left, 130 bottom, 150, 162 bottom; **Agusta** 163 bottom; **Airship Industries** 23 top; **Bell Helicopter Textron** 45 bottom, 166 top, 166 bottom; **Boeing** 167 bottom right; **Brian M. Service** Title spread, 41 bottom, 173 top, 178; **British Aerospace** 51 top, 80, 102 top, 110–111, 134 bottom, 135, 136, 138 middle, 139 top, 139 bottom, 143, 145, 167 top; **British Aerospace Dynamics** 58, 128 middle, 130 top; **Canadair** 170 top; **Dassault-Breguet** 71 top, 79 bottom; **De Havilland Canada** 170 middle, 171 top; **ECP/Armées** 163 top, 167 bottom left, 186; **Embraer** 171 middle; **Ferranti Radar Systems** 146; **Fleet Air Arm** 161; **Fleet Air Arm Museum** 40 bottom; **G. Arra** 152 bottom; **GEC Avionics** 70 bottom, 78 top; **General Dynamics** 43 top; **General Electric** 112, 114; **Government Aircraft Factories (GAF), Australia** 79 top; **Grumman Aerospace Corporation** 34–35 bottom, 46–47, 54 middle, 86–87, 95 top, 95 left, 98 bottom left, 102 middle, 107 top, 122 top, 171 bottom, 173 bottom; **Hughes Aircraft** 56, 94 middle, 106, 117, 121 bottom; 127 top; **Imperial War Museum** 18 top, 19, 22, 24 middle, 24 bottom, 35 top right; **Jane's Defence Weekly** 151, 152 top; **Kaman Aerospace** 175; **Kongsberg Vaapenfabrikk** 62; **Kongsberg Vaapenfabrikk** 62, 126 top right, 127 bottom; **Lockheed-California** 74 top, 74 bottom; **LTV** 186–187; **Marine Nationale** 162 top; **McDonnell Douglas** 7, 44, 51 bottom, 82 top, 90 top, 91, 107 middle, 109 bottom, 115, 116 bottom, 128 top, 137; **Messerschmitt Bolkow-Blöhm** 126 bottom, 182 bottom; **Ministry of Defence** 38 bottom; **National Maritime Museum, Greenwich** 10; **Pilot Press** 14–15, 28–29, 32–33, 36–37, 52–53, 76–77, 92–93, 96–97, 124–125, 140–141; **Plessey Aerospace** 128 bottom, 129 bottom; **Plessey Marine** 75; **Raytheon Company** 94 top; **Rockwell International** 182 top; **Rolls-Royce** 134 middle, 147; **Royal Air Force** 113 top; **Royal Danish Air Force** 72, 129 top, 185; **Royal Navy** 57, 116 top, 132–133, 148–149, 157; **Short Brothers** 13 right, 17; **Sikorsky Aircraft** 78 bottom, 85, 176–177; **Swan Hunter** 154; **Swedish Air Force** Half-title, 71 bottom, 83 bottom; **Target Technology Ltd** 118; **The Plessey Company PLC** 123; **Thorn/EMI** 73 top; **US Air Force** 38 top, 45 top; **US Department of Defense** 49, 50, 55 bottom, 59, 60–61 top, 60–61 bottom, 62–63 bottom, 66–67, 68, 82 bottom, 131; **US National Archives** 11, 12, 27 top, 27 middle, 31 bottom; **US Navy** 8–9, 16, 20 top, 20 bottom, 21, 26 top, 26 middle, 34–35 top, 39 top, 39 bottom, 42, 43 bottom, 54 top, 64, 64–65 top, 65, 90 middle, 99, 113 bottom, 120 top, 155, 156, 172.